Modern Tests for 'A' Level Chemistry

Second Edition

Alan J. Furse

Edward Arnold

© Alan J. Furse 1972, 1985

First published in Great Britain 1985
by Edward Arnold (Publishers) Ltd
41 Bedford Square
London WC2 3DQ

Edward Arnold (Australia) Pty Ltd
80 Waverley Road
Caulfield East 3145
PO Box 234
Melbourne

British Library Cataloguing in Publication Data
Furse, Alan J.
  Modern tests for 'A' Level chemistry.
  —2nd ed.
  1. Chemistry—Examinations, questions, etc.
  I. Title
  540'.76   QD42

ISBN 0 7131 7314 9

All rights reserved. No part of this publication may be reproduced, stored in a retrieval system, or transmitted in any form or by any means, electronic, photocopying, recording, or otherwise, without the prior permission of Edward Arnold (Publishers) Ltd

Text filmset in 9/10 Univers
by CK Typesetters Ltd, Sutton, Surrey
Printed and bound by
Thomson Litho Ltd, East Kilbride, Scotland

Contents

Types of question
1	The Periodic Table 1: an introduction	1
2	The Periodic Table 2: the elements of Groups I and II	5
3	The properties of gases	8
4	Atomic structure	14
5	The Periodic Table 3: the elements of Group VII	19
6	Energy changes and bonding	23
7	Crystal structure	29
8	Chemical bonding	32
9	Organic chemistry 1: hydrocarbons and halogenoalkanes	35
10	Intermolecular forces	40
11	Organic chemistry 2: —OH group compounds	43
12	Equilibria	48
13	Organic chemistry 3: amines, carbonyl compounds, proteins	54
14	Rate of reaction	61
15	Voltaic cells	67
16	The Periodic Table 4: the transition elements	72
17	Organic chemistry 4	79
18	The Periodic Table 5: some p-block elements	84

Preface

This is a collection of modern test questions in 'A' Level Chemistry. It is intended primarily to accompany the 'A' Level Chemistry course as produced by Nuffield Advanced Science and revised by the Nuffield Chelsea Curriculum Trust.

There are 18 sections in the book, corresponding approximately to the 18 Topics of the Nuffield course. Each section contains fixed response items and structured questions with a suggested mark distribution.

Within each section the questions tend to be relevant only to one particular Topic, though inevitably there is some increase in sophistication as the subject develops. The sections are thus intended for use as tests at the ends of Topics and for this purpose are preferable to past examination questions which often contain material from more than one Topic and are appropriate in sophistication for those who have completed the course.

Some of this material is adapted from the first edition of 'Modern Tests' but much of it is new.

My thanks are due to my colleagues at Blundell's, Barry Wood and Ian Maines, for helpful criticism and to their students and mine for using the questions and providing much vigorous discussion of mutual benefit.

AJF

Types of question

Three types of fixed response item have been used, namely (a) multiple choice, (b) classification and (c) multiple completion. These three types are mixed together throughout the text but each type is readily identifiable by its format.

(a) Multiple choice: choose the **one** correct, or most appropriate, answer from the five lettered suggestions.
(b) Classification: choose the **one** category, diagram, equation etc. from those offered which best fits the information in the question.
(c) Multiple completion: three choices are given, marked **1** to **3**, which may be appropriate answers to, or completions of, the question. Select the ones which are correct and answer as follows:

    A if **1, 2** and **3** are all correct
    B if **1** and **2** only are correct
    C if **2** and **3** only are correct
    D if **1** only is correct
    E if **3** only is correct

SECTION 1

The Periodic Table 1: an introduction

1A Fixed response items

1. As the first stage in the preparation of some magnesium compounds it is necessary to react 10 g of magnesium oxide with dilute sulphuric acid. What is the minimum volume of 2 M sulphuric acid which must be used? (MgO = 40)

   A 125 cm^3 B 75 cm^3 C 150 cm^3 D 250 cm^3 E 100 cm^3

2. What volume of 0.4 M sodium hydroxide solution could be made from 20 g of solid sodium hydroxide? (NaOH = 40)

   A 500 cm^3 B 5 dm^3 C 125 cm^3 D 1.5 dm^3 E 1250 cm^3

3. How many moles of water of crystallization are there in 0.2 mole of sodium sulphate $Na_2SO_4 \cdot 10H_2O$?

   A 0.2   B 2   C 5   D 10   E 20

4. When 1.11 g of calcium chloride $CaCl_2$ is dissolved in water and the solution is made up to 100 cm^3, which of the following represents the concentration of the chloride ions? ($CaCl_2$ = 111)

   A 0.2 M   B 0.02 M   C 0.1 M   D 0.05 M   E 0.01 M

5. Each of the following carbonates decomposes on heating and in each case the products are the oxide of the metal and carbon dioxide. If 1 g of each carbonate is decomposed, which of them produces the greatest volume of carbon dioxide? (Ca = 40, Li = 9, Mg = 24, Cu = 64, Zn = 65)

   A $CaCO_3$ B Li_2CO_3 C $MgCO_3$ D $CuCO_3$ E $ZnCO_3$

6. 1 g of chlorine is allowed to react completely with each of the metals magnesium, calcium, barium, sodium and potassium. The formulae of the chlorides are respectively $MgCl_2$, $CaCl_2$, $BaCl_2$, NaCl and KCl. From which of the metals will the greatest mass of chloride be produced? (K = 39, Ba = 137)

   A Mg   B Ca   C Ba   D Na   E K

7. When 0.4 mole of aluminium sulphate $Al_2(SO_4)_3$ is dissolved in water and the volume of the solution is made up to 500 cm^3

   1 1.2 moles of sulphate ions are in solution
   2 a total of 2 moles of ions are in solution
   3 the solution is 1.6 M $Al_2(SO_4)_3$

8-12 In each of these items, choose the answer from the following list:

A 0.05   **B** 0.4   **C** 1.6   **D** 2.4   **E** 3.2

8 The number of moles of hydrogen atoms in 0.4 moles of $LiAlH_4$

9 The number of moles of Cl^- ions in 0.2 moles of $CaCl_2$

10 The number of moles of chlorine atoms in 0.6 moles of CCl_4

11 The number of moles of nitrate ions NO_3^- in 250 cm³ of 0.2 M $AgNO_3$

12 The concentration in mol dm⁻³ of a solution of silver nitrate $AgNO_3$, 20 cm³ of which react exactly with 5 cm³ of 0.8 M $CaCl_2$

($Ag^+ + Cl^- \rightarrow AgCl$)

13-16 These items are concerned with the method of determining the Avogadro constant which depends on radioactivity.

13 Which of the following is a suitable radioactive element to use?

A carbon   **B** potassium   **C** nitrogen
D technetium   **E** radium

14 If a radioactive element has a count rate of n alpha particles per second, how many counts will be produced in a year?

A $n \times 60 \times 365$
B $n \times 365$
C $n \times 60 \times 60 \times 24 \times 365$
D $n/(60 \times 60 \times 24 \times 365)$
E $\dfrac{365n}{60 \times 60 \times 24}$

15 If a radioactive sample gives a total of N counts per year and in this time V cm³ of helium are collected, measured at s.t.p., which of the following gives the Avogadro constant?

A $\dfrac{N}{V \times 22.4}$ **B** $\dfrac{N \times 22\,400}{V}$ **C** $\dfrac{NV}{22.4}$

E $\dfrac{NV}{22\,400}$ **E** $NV \times 22\,400$

16 Which of the following quantities is the one whose measurement is most likely to be responsible for inaccuracies in this method?

A the count rate of radioactive material
B the time over which the helium is collected
C the volume of helium collected
D the temperature of the helium collected
E the volume of 1 mole of helium

7 Elements in the same Group of the Periodic Table

1. have the same atomic number
2. have similar formulae for their compounds
3. increase in atomic radius as one goes down the group

8 The transition metals

1. occur between Group II and Group III of the Periodic Table
2. tend to have ions with a variable number of charges on them
3. tend to form coloured compounds

9 Elements towards the left-hand end of each period of the Periodic Table

1. tend to be metals
2. tend to have lower melting points than those at the right-hand end
3. have a greater tendency to combine with metals than with non-metals

20 True statements about the elements of the 3rd Period of the Periodic Table (Na to Ar) taken in order, include:

1. their melting points rise to a maximum at silicon and then fall gradually to the right-hand end
2. the boiling points of the elements rise to a maximum at aluminium, then drop sharply
3. the heats of vaporization of the elements rise to a maximum at silicon, then drop sharply

1B Structured questions

1 The short-answer questions which follow are about this incomplete form of the Periodic Table:

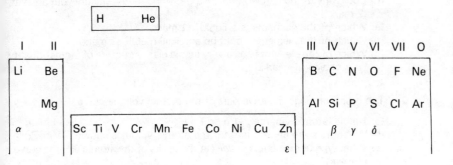

You will have noticed that five elements in this table have been given Greek letters, which are obviously not their chemical symbols.

(a) Which of the elements α–ε would react with cold water? (1 mark)
(b) Which of the elements α–ε is most non-metallic? Give a reason for your choice. (2 marks)

(c) Which of the five elements is most likely to be intermediate in properties between a metal and a non-metal? (1 mark)
(d) Which of the five elements is a transition metal? (1 mark)
(e) Which pair, chosen from the five elements, would you expect to react together with the evolution of the most heat per mole of formulae of product? (2 marks)
(f) Which of the five elements is least likely to form ionic binary compounds? (1 mark)
(g) Which of the five elements would form a compound with hydrogen having 1 atom of hydrogen per formula of the hydride? (1 mark)
(h) Which of the five elements would form an oxide having 2 moles of the element for every 1 mole of oxygen? (1 mark)

2   The following information is about *six* elements which are consecutive in atomic number in the periodic table.

element	m.p. °C	ΔH_{fus} kJ mol^{-1}	density g cm^{-3}
a	1410	46.63	2.42
b	44.3	0.63	1.83
c	118.9	1.68	2.07
d	−101.0	3.36	0.002 99
e	−189.4	1.26	0.001 66
f	63.4	2.52	0.87

(a) Which of these six elements have molecular structures? Justify your choice. (2 marks)
(b) Why are the values of two of the densities so much lower than the other values? (1 mark)
(c) To what extent is there a relationship between the figures in the melting point column and those in the heat of fusion column? (1 mark)
(d) Could any of the six elements be a transition metal? Justify your answer. (2 marks)
(e) Which of the elements is a liquid? (1 mark)
(f) Which of the elements could be an alkali metal? (1 mark)
(g) Which of the elements has a very stable giant structure of atoms? Justify your choice. (2 marks)

3   The graph on page 5 shows part of the atomic volume curve.

(a) Define 'atomic volume'. (1 mark)
(b) What units are appropriate for the atomic volumes shown? (1 mark)
(c) What type of element is *usually* at the peaks of the atomic volume curve? (1 mark)
(d) One of the peaks is unusual; which one and why? (2 marks)
(e) What type of element is in the region marked X? (1 mark)
(f) Why do the gaseous elements not have *much* higher atomic volumes than the solid elements on this curve? (2 marks)
(g) Without using a data book, *estimate* the density of element Y, stating clearly what assumptions you make. (2 marks)

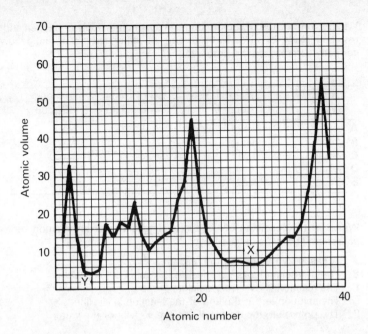

SECTION 2

The Periodic Table 2: the elements of Groups I and II

2A Fixed response items

1. When a certain solid substance Z was heated strongly in a test tube, the gaseous contents of the test tube were found to be acidic and to rekindle a glowing spill. The residue was found to colour a bunsen burner flame green. Which of the following could Z have been?

   A strontium nitrate **B** barium nitrate **C** copper carbonate
   D barium carbonate **E** potassium carbonate

2. Which of the following nitrates is hydrated?

   A sodium nitrate **B** potassium nitrate **C** magnesium nitrate
   D strontium nitrate **E** barium nitrate

3. Rubidium chloride would be expected to be

   1 hydrated
   2 ionic
   3 soluble in water

5

4 When a 3 mm cube of sodium burns in a plentiful supply of oxygen and the resulting oxide is mixed with 10 cm³ of water, which of the following correctly summarizes the observations?

	flame colour	state of oxide (room temp.)	pH of oxide/ water mixture	formula of oxide
A	yellow	solid	14	Na_2O_2
B	green	liquid	8	Na_2O_2
C	yellow	solid	6	Na_2O
D	yellow	solid	4	Na_2O
E	green	gas	10	Na_2O_2

5 Which of the following observations suggests that the reaction between potassium and water is exothermic?

A A gas is formed during the reaction.
B The potassium looks spherical as it reacts with the water.
C The reaction is over very quickly.
D The solution left at the end of the reaction is alkaline.
E The potassium moves rapidly on the surface of the water.

6 When metal ions in solution react with a cation exchange resin in its hydrogen form

1 positive ions are replaced by negative ions
2 one metal ion replaces one hydrogen ion from the resin
3 the solution coming through the resin becomes more acidic

7-10 These items concern the following experiment. About 0.153 g (0.001 mol) of barium oxide was placed in a test tube and 5 cm³ of M hydrochloric acid (0.005 mol) was added. On shaking the test tube the solid dissolved. On adding 2 cm³ of M sulphuric acid to the mixture there was a white precipitate.

7 If 5 cm³ of water had been used instead of the M hydrochloric acid

A the solid would have dissolved easily
B the resulting mixture would have been acidic
C a substance of formula BaOH would have been produced
D there would have been effervescence
E barium hydroxide would have been formed

8 After adding the hydrochloric acid the mixture

A turned green
B was alkaline
C contained some $BaCl_3$
D was neutral
E contained excess hydrochloric acid

The precipitate mentioned in the description was

A barium oxide
B barium sulphate
C barium chloride and barium hydroxide
D barium hydroxide
E barium chloride and barium sulphate

10 If 5 cm³ of M nitric acid had been used instead of the M hydrochloric acid

A there would have been no change in the observations
B barium nitrate would have been precipitated
C there would have been no precipitate
D the solid barium oxide would not have dissolved
E there would have been effervescence on adding the M sulphuric acid

2B Structured questions

1 Give an example of each of the following, mentioning both name and formula when you refer to a compound:

(a) a soluble carbonate (more than 0.1 mol dm⁻³) (1 mark)
(b) an insoluble sulphate (less than 1×10^{-3} mol dm⁻³) (1 mark)
(c) a metal ion which gives a purplish-pink flame colour (1 mark)
(d) an anhydrous nitrate (1 mark)
(e) a carbonate which does not decompose into the oxide and carbon dioxide when heated in a Bunsen flame (1 mark)
(f) a use for ion exchange (1 mark)
(g) the difference between the physical properties expected from a Group II and a Group VII element (2 marks)
(h) a way in which a compound of lithium resembles compounds of Group II elements rather than those of Group I elements. (2 marks)

2 Some white crystals put out in a petri dish for an experiment quickly became damp-looking (a). On heating in a test tube they became liquid almost at once (b) and then the liquid boiled (c). After a time the contents of the test tube became a white solid (d). On further heating a coloured gas (e) and oxygen were evolved. The residue was a white solid which when mixed with water gave a strongly alkaline mixture (f). A little of the final residue was put on to a flame test wire moistened with concentrated hydrochloric acid. An orange-red flame was produced which had dull-red edges (g).

Identify the original crystals and write a sentence of explanation of each of the observations marked (a) to (g). (1 mark for each of (a)–(e); (f) 2 marks; (g) 1 mark; identity 2 marks)

3 A certain mass of a solid mixture of strontium chloride $SrCl_2$ and strontium hydroxide $Sr(OH)_2$ was dissolved in water and the solution made up to exactly 100 cm³ with water. 10 cm³ of this solution was titrated with 0.001 M HCl and 19.0 cm³ was required for neutralization. A second 10 cm³ portion of the solution was washed through a strong cation exchange resin in its hydrogen form and the washings required 13.5 cm³ of 0.001 M NaOH for neutralization.

(a) What indicator would be used for the first titration (1 mark)
(b) How many moles of H^+ are there in 19.0 cm³ of 0.001 M HCl? (1 mark)

(c) How many moles of OH⁻ did these react with? (1 mark)
(d) How many moles of $Sr(OH)_2$ were there in the 10 cm³ portion of the solution taken? (1 mark)
(e) How many moles of $Sr(OH)_2$ were there in the mixture at the start? (mark)
(f) What happened to the OH⁻ ions in the mixture during the ion exchange (2 marks)
(g) Calculate the number of moles of $SrCl_2$ in the mixture at the start, using the second titration result. (3 marks)

SECTION 3

The properties of gases

3A Fixed response items

In this section R is used in place of Lk, where L is the Avogadro constant and k the Boltzman constant.

1. Which of the following gaseous compounds is more dense than air at room temperature and pressure?

   A C_2H_2 **B** C_2H_4 **C** CH_4 **D** CH_3NH_2 **E** NH_3

2. In a determination of the relative molecular mass of a liquid 0.2 cm³ of the liquid (density 1.50 g cm⁻³) was found to give 56.3 cm³ of vapour (corrected to s.t.p.). What is the relative molecular mass of the liquid?

   A 79.6 g   **B** 51.7 g   **C** 119.4 g   **D** 159.2 g   **E** 103.4 g

3. A certain compound containing carbon and hydrogen only is analysed and found to contain 92.3% of carbon. 0.195 g of the compound was vaporized and found to occupy 56 cm³ (corrected to s.t.p.). What is the formula of the compound?

   A C_6H_{12} **B** C_4H_4 **C** C_2H_2 **D** C_6H_6 **E** C_6H_{10}

4. The units of the Avogadro constant are

   A mol⁻¹   **B** dm³   **C** mol   **D** dm⁻³   **E** g⁻¹

5. When 70 cm³ of an oxide of nitrogen is reacted with heated sulphur the residual volume of gas is 105 cm³, which on addition of excess aqueous alkali reduces to 70 cm³, all measurements being made at the same temperature and pressure. Which of the following equations is consistent with these figures?

   A $3N_2O(g) + S(s) \rightarrow SO_3(s) + 3N_2(g)$
   B $2N_2O(g) + S(s) \rightarrow SO_2(g) + 2N_2(g)$

C $2NO_2(g) + 2S(s) \rightarrow 2SO_2(g) + N_2(g)$
D $2NO(g) + S(s) \rightarrow SO_2(g) + N_2(g)$
E $6NO(g) + 2S(s) \rightarrow 2SO_3(s) + 3N_2(g)$

6 If p is the density of a gas, M its relative molecular mass, R is the gas constant, T the temperature and P its pressure, which of the following statements is true?

A $p = \dfrac{RT}{PM}$ B $p = \dfrac{RM}{PT}$ C $p = \dfrac{RMP}{T}$

D $p = \dfrac{RTM}{P}$ E $p = \dfrac{PM}{RT}$

7 If a certain mass of gas occupies 65 cm³ at 293 K and 770 mmHg, what is its volume at s.t.p.?

A $\dfrac{1}{65} \times \dfrac{273}{293} \times \dfrac{770}{760}$ cm³   B $65 \times \dfrac{293}{273} \times \dfrac{770}{760}$ cm³

C $65 \times \dfrac{273}{293} \times \dfrac{760}{770}$ cm³   D $65 \times \dfrac{273}{293} \times \dfrac{770}{760}$ cm³

E $\dfrac{1}{65} \times \dfrac{273}{293} \times \dfrac{770}{760}$ cm³

8 Which of the following reactions is accompanied by the largest percentage increase in volume?

A $N_2H_4(l) + 2H_2O_2(l) \rightarrow N_2(g) + 4H_2O(g)$
B $2NH_3(g) \rightarrow N_2(g) + 3H_2(g)$
C $2H_2O_2(l) \rightarrow 2H_2O(l) + O_2(g)$
D $MnO_2(s) + 4HCl(g) \rightarrow 2H_2O(g) + Cl_2(g) + MnCl_2(s)$
E $2Al(s) + 3H_2SO_4 \text{ [0.1 M (aq)]} \rightarrow Al_2(SO_4)_3(aq) + 3H_2(g)$

9 The mass of a certain element in 1 mole of each of its gaseous compounds was found always to be one of the following (in grams): 19, 38, 57, 76, 95, 114.
Which of the following is a possible value for the relative atomic mass of the element?

A 9.5   B 19   C 28.5   D 38   E 114

10 Which of the following statements about Avogadro's theory and its consequences is *un*true?

A Equal volumes of hydrogen and of sulphur dioxide contain the same number of molecules at s.t.p.
B Two volumes of hydrogen and one volume of methane CH_4 contain equal numbers of hydrogen atoms at s.t.p.
C Equal volumes of carbon monoxide and of methane contain equal numbers of carbon atoms at s.t.p.
D The volume occupied by three moles of chlorine atoms is approximately 33.6 dm³ at s.t.p.
E The volume occupied by 1 mole of atoms of any gaseous element is approximately 11.2 dm³ at s.t.p.

11 A pencil line is assumed to be made of graphite and is t cm thick. It is 0.05 cm wide and 10 cm long. If the density of carbon is 2.25 g cm^{-1} and the Avogadro constant is L, how many carbon atoms are there in the line? (C = 12)

A $\dfrac{0.05 \times 10 \times 2.25 \times L \times t}{12}$ **B** $\dfrac{0.05 \times 10 \times L \times t}{12 \times 2.25}$

C $\dfrac{12 \times L \times t}{0.05 \times 10 \times 2.25}$ **D** $\dfrac{12 \times 2.25 \times L \times t}{0.05 \times 10}$

E $\dfrac{L \times t}{0.05 \times 10 \times 12 \times 2.25}$

12 One 'Faraday' (96 500 C) is sometimes called '1 mole of electricity' because

1. it is the charge on 1 mole of electrons
2. it is the charge on the electrons in 1 mole of atoms of an element
3. it is the charge on the electrons in 1 mole of carbon atoms

13 A plastic gas syringe was weighed with 50 cm³ of dry air in it (converted to s.t.p. this was 46 cm³). It was weighed again with 50 cm³ of a gas G in it. The results were:

    mass of syringe + air = 45.259 g
    mass of syringe + G = 45.295 g
    mass of air in syringe = 0.059 g

What is the relative molecular mass of G?

A $\dfrac{0.095 \times 22\,400}{46}$ **B** $\dfrac{0.095 \times 22\,400}{50}$

C $\dfrac{0.036 \times 22\,400}{46}$ **D** $\dfrac{0.036 \times 22\,400}{50}$

E $\dfrac{0.059 \times 22\,400}{46}$

14-16 These items concern the following account of an experiment to find the relative molecular mass of a volatile liquid by injecting a known mass of it into a glass gas syringe which is being heated by passing steam round it. 0.375 g of the liquid was found to produce 52 cm³ of vapour after correction to s.t.p.

14 Which of the following pieces of information would *not* be required for obtaining the volume at s.t.p.?

A the pressure of the atmosphere
B the temperature of the laboratory
C the temperature of the steam jacket
D the temperature of melting ice
E the pressure of one standard atmosphere

15 Which of the following gives the mass of 1 mole of the liquid in grams?

A $\dfrac{0.375 \times 52}{22\,400}$ B $0.375 \times 52 \times 22\,400$

C $\dfrac{0.375 \times 22\,400}{52}$ D $\dfrac{52 \times 22\,400}{0.375}$

E $\dfrac{22\,400}{52 \times 0.375}$

16 Which of the following errors would tend to make the measured value of the relative molecular mass *smaller* than it should be?
A loss of liquid into the atmosphere during injection
B failure of all the liquid to vaporize inside the gas syringe
C having more than 10 cm³ of air in the gas syringe before injection
D estimating the temperature of the steam bath as 383 K instead of the correct value of 373 K
E injecting a substantial quantity of air along with the volatile liquid

17 If a certain mass of gas has a volume of 1 dm³ and then its volume is doubled
A its pressure and entropy both increase
B its entropy doubles
C its pressure and entropy both decrease
D its entropy halves
E its pressure decreases but its entropy increases

18 In the expression $pV = \tfrac{1}{3} Lmu^2$ for 1 mole of an ideal gas
1 at constant temperature pV would be expected to be constant
2 for a given gas, L and m are constant
3 at constant pressure, the volume of a gas would be inversely proportional to its temperature

19 Which of the following is likely to have the highest standard entropy?
A $O_2(g)$ B $H_2O(l)$ C $C_2H_5OH(l)$
D $H_2O(s)$ E $C_6H_{12}O_6(s)$

20 Which of the following are the units of standard entropy change?
A J K mol   B J K^{-1} mol   C J^{-1} K mol^{-1}
D J^{-1} K^{-1} mol   E J K^{-1} mol^{-1}

3B Structured questions

1 Caesium chloride has a structure in which eight chloride ions are at the corners of a cube with a caesium ion at the centre of the cube thus:

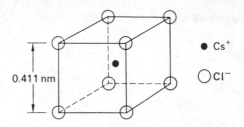

Other data for caesium chloride are:

mass of 1 mole caesium chloride CaCl = 168.36 g
density of caesium chloride = 3.998 g cm^{-3}

(a) How many caesium ions are totally and exclusively within the unit cube? (2 marks)
(b) How many chloride ions are exclusively within the unit cube? (2 marks)
(c) What is the volume of a unit cube of caesium chloride? (2 marks)
(d) What is the volume occupied by 1 mole of caesium chloride crystals CsCl? (2 marks)
(e) Use your answers to (c) and (d) to calcuate a value of the Avogadro constant L. (2 marks)

2   It has been suggested[1] that a rough estimate of the relative molecular mass of carbon dioxide can be made by rapidly weighing a piece of solid carbon dioxide and then allowing it to evaporate into a gas syringe. Solid carbon dioxide does not melt at atmospheric pressures, it changes from the solid state directly into a gas. At 760 mmHg the temperature at which this happens is −78°C.

(a) Draw a diagram of the apparatus you would use to 'allow it [the carbon dioxide] to evaporate into a gas syringe'. (2 marks)
(b) Suggest a suitable method for weighing the solid carbon dioxide rapidly, bearing in mind that it has to be transferred to the apparatus you have designed very quickly. (3 marks)
(c) Most gas syringes in use in school laboratories have a graduated volume of 100 cm³. Calculate to the nearest 0.01 g the maximum mass of solid carbon dioxide which ought to be used in this experiment. (2 marks)
(d) In this experiment, there are a number of sources of error, some peculiar to the experiment, others inherent in the use of the relationship '1 mole of a gas at s.t.p. occupies 22 400 cm³'. List the principal sources of error in what seems to you to be their order of importance, beginning with the most important. (3 marks)

3   It has been reported[2] that scientists working at the Sydney laboratories of the Commonwealth Division of Mineral Chemistry have developed a new way of obtaining ethyne (C_2H_2) from natural gas (methane, CH_4) by a process called 'electro-cracking'. This process involves passing an electric discharge through methane at a pressure of about 76 mmHg. The report

states that over 80% of the methane is converted into ethyne in a single passage through the discharge and that about 3 cubic feet of hydrogen are produced for each cubic foot of ethyne. Electro-cracking can also be used to convert benzene vapour (C_6H_6) into ethyne.

(a) Write an equation for the conversion of methane into ethyne by electro-cracking. (2 marks)
(b) From the equation, calculate the maximum volume of hydrogen which could be produced from each cubic decimetre of methane assuming the process to be 80% efficient. (2 marks)
(c) Is your answer consistent with the figures given in the report? Does it make any difference that the volume units used are different? (2 marks)
(d) What assumptions are made about temperature and pressure in the calculation in question (b)? (2 marks)
(e) Calculate the maximum volume of ethyne which can be produced from 1 mole of liquid benzene, assuming that at 298 K and 760 mmHg, the molar volume is 24.2 dm³ and that the process is 80% efficient. (2 marks)

4   The apparatus illustrated below has been used[3] to determine the relative molecular masses of gases.

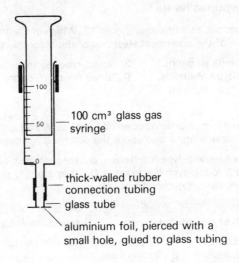

Method

The syringe is filled with the gas to be studied, the glass tube is fitted on the nozzle and the plunger is allowed to fall under its own weight, expelling gas through the small hole in the foil. The time taken for 100 cm³ of gas to be expelled is measured with a stop-clock.

Theory

According to the law of effusion:

Time taken to expel 100 cm³ of gas $\propto \sqrt{\text{relative molecular mass of the gas}}$

(a) The expression given above under 'theory' is a proportionality. What experiments would you do and what assumptions would you make in order to determine the relative molecular mass of a gas whose chemical nature you did not know? (4 marks)

(b) When the glass tube is fitted into position it contains air, so the first gas to be expelled from the syringe is air. Does this introduce a significant error into the experiment? Justify your answer. (2 marks)

(c) Would better results be obtained by using a larger or a smaller hole? Justify your answer. (2 marks)

(d) Arrange the following in increasing order of time taken to expel 100 cm³: hydrogen bromide, bromine vapour, carbon dioxide, nitrogen, air, argon. (H = 1, Br = 80, Ar = 40, C = 12, O = 16, N = 14) (2 marks)

SECTION 4

Atomic structure

4A Fixed response items

1 Magnesium has an atomic number of 12. Which of the following is the best description of the outermost electrons of the magnesium atom?

   A two s-type electrons   **B** one s-type electron
   C two p-type electrons   **D** three p-type electrons
   E one p-type electron

2 When aluminium (atomic number 13) forms its most usual ions, which of the following statements is true about the outermost electrons of these *ions*?

   A they are eight p-type electrons   **B** they are six p-type electrons
   C they are four p-type electrons   **D** they are two p-type electrons
   E they are two s-type electrons

3 The atoms of sodium (atomic no. = 11) have as their outermost electrons:

   A 1s electrons   **B** 2s electrons   **C** 3s electrons
   D 2p electrons   **E** 3p electrons

4 Which of the following types of electrons do not exist?

   A 2p electrons   **B** 5s electrons   **C** 5d electrons
   D 4s electrons   **E** 2d electrons

5 Atoms of elements in the first transition series of metals differ from each other *mainly* in the number of

   A s electrons   **B** p electrons   **C** s and p electrons
   D d electrons   **E** p and d electrons

6 The atoms of isotopes of the same element differ from each other in the numbers of

- **A** s electrons
- **B** neutrons
- **C** p electrons
- **D** protons
- **E** d electrons

7 Atoms of isotopes of the same element are identical in which of the following respects?

- **A** Atomic structures
- **B** Numbers of electrons outside the nucleus
- **C** The sum of the number of protons and the number of neutrons
- **D** The sum of the number of electrons and the number of neutrons
- **E** Nuclear masses

8 For which of the following elements would you expect it to require the least energy to remove one electron from a gaseous atom?

- **A** argon
- **B** aluminium
- **C** chlorine
- **D** lithium
- **E** nitrogen

9 A certain element Y has an atomic number twice that of an element X. The first three ionization energies for these elements are as follows:

	1st	2nd	3rd	
X	528	7340	11850	kJ mol^{-1}
Y	1095	2370	4660	

Which of the following could be the identity of X and Y? (Numbers in brackets are atomic numbers)

- **A** beryllium (4) and oxygen (8)
- **B** lithium (3) and carbon (6)
- **B** helium (2) and beryllium (4)
- **D** boron (5) and neon (10)
- **E** carbon (6) and magnesium (12)

10 A certain element has a relative atomic mass approximately $(2x + 10)$ where x is its atomic number. Which of the following is the number of neutrons in the nucleus of this atom?

- **A** $x + 10$
- **B** $2x$
- **C** 10
- **D** x
- **E** $x + 5$

11 Which of the following best represents the change accompanied by the energy term we call the '1st ionization energy of the element X'?

- **A** $X(s) = X^+(g) + e^-$
- **B** $X(g) = X^+(s) + e^-$
- **C** $X^-(g) = X(g) + e^-$
- **D** $X^+(g) = X^{2+}(g) + e^-$
- **E** $X(g) = X^+(g) + e^-$

12-17 These items each refer to the following sketches of graphs, all of which have *energy* on the vertical axis.

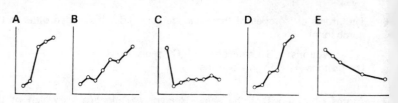

12 Which of the diagrams could be a graph of 1st ionization energy of the noble gases, He, Ne, Ar, etc. plotted against atomic number?

13 Which of the diagrams could represent a graph of 1st ionization energies for a series of elements, consecutive in atomic number and beginning with a noble gas?

14 Which of the diagrams could represent the 1st, 2nd, 3rd etc. ionization energies for the element magnesium, plotted against number of electrons removed?

15 Which of the diagrams could represent the 1st ionization energies of a 'short period' of the Periodic Table plotted against atomic number?

16 Which of the diagrams could represent the 1st, 2nd, 3rd etc. ionization energies for the element carbon (atomic number 6), plotted against number of electrons removed?

17 Which of the diagrams could represent the 1st, 2nd, 3rd etc. ionization energies of the element calcium, plotted against number of electrons removed?

18 In Geiger and Marsden's α-particle scattering experiment
 A more than 90% of α-particles were deflected
 B electrons were responsible for the deflection of the α-particles
 C α-particles frequently collided with the nuclei of the atoms in the foil
 D the positive nuclei of the atoms repelled α-particles which approached them
 E the α-particles changed the chemical nature of the atoms in the foil

19 In an experiment to measure the 1st ionization energy of an element, an accelerating potential of 16.2 volts is applied between electrodes placed in the vapour of the element in order *just* to produce ionization. Data: electronic charge is 1.6×10^{-19} C; L is 6.2×10^{23}; 1 kJ is 1000 J. Which of the following gives the 1st ionization energy of the element?

 A $16.2 \times 1.6 \times 6.2 \times 10$
 B $\dfrac{16.2 \times 1.6}{6.2}$
 C $\dfrac{16.2 \times 6.2 \times 10^7}{1.6}$
 D $\dfrac{16.2 \times 1.6 \times 10^{-7}}{6.2}$
 E $\dfrac{16.2 \times 6.2 \times 10}{1.6}$

20 A particular element has the ground state electronic configuration

$1s^2\ 2s^2\ 2p^6\ 3s^2\ 3p^6\ 4s^2\ 3d^6$

This element

1 has an atomic number of 26
2 is a transition element
3 is in Group VI of the Periodic Table

4B Structured questions

1 The following diagram shows a mass spectrometer trace for the element magnesium, Mg (atomic no. = 12).

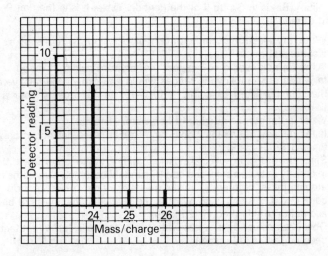

(a) How many (i) protons and (ii) neutrons are there in each nucleus of the isotope of mass 26? (2 marks)
(b) When the particles are ionized in the mass spectrometer, what is the formula for the ions which give rise to the above trace? (1 mark)
(c) Indicate by simple sketches of the type below what happens to the ions of masses 24 and 26 when those of type 25 are landing on the detector of the mass spectrometer. (2 marks)

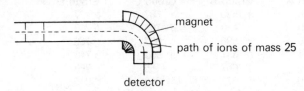

(d) Using the data on the trace, calculate a value for the relative atomic mass of magnesium.
(e) Why is this value still not as accurate as the best value given in the Book of Data (24.3120)? (2 marks)

2   The questions which follow are related to the subject matter of this article:[4]

Beryllium, used for some time in ballistic missiles and spacecraft because of its heat absorption and dissipation properties, is beginning to be employed in certain aircraft parts. Brake discs made of this metal took the place of steel discs in the production version of the Lockheed C5 transport aeroplane for the USAF. Apart from its heat resisting characteristics (five times better than those of steel) it also affords a weight advantage. One steel brake disc of the type used in the C5 weighs as much as four of the same type made in beryllium. On the big undercarriage of the C5, the change to beryllium for the brake discs means a saving of 1800 lb. The higher cost is expected to be offset by longer working life.

Use of beryllium has been increasing since about 1950 when it was employed to reduce the velocity of neutrons in nuclear reactors. It has since found a wide application in the heat shields of missiles and satellites.

Beryllium (Be) is in Group II of the Periodic Table. It is in the first Period of eight elements.

(a) Write the electronic configuration of beryllium in terms of s, p etc. electrons (1 mark)
(b) What formula would you expect for beryllium oxide? (1 mark)
(c) In which of the following ranges would you expect the relative atomic mass of beryllium to come: 1–5, 6–10, 11–15? Justify your answer. (2 marks)
(d) What is the atomic number of beryllium? (1 mark)
(e) If the density of steel is 7.86 g cm^{-3}, what is the density of beryllium? (lines 8–9). (1 mark)
(f) Since the atoms in solids are packed fairly close together, the difference in density between steel and beryllium could be largely due to the relative atomic mass difference. If the relative atomic mass of iron (major component of steel) is 56, what is the relative atomic mass of beryllium according to this theory? (2 marks)
(g) Does the answer to (f) agree with your estimate in (c)? What factor has been neglected? (2 marks)

3   The 1st ionization energies (in kJ mol^{-1}) for a number of elements are given below. The elements are arranged in four groups, each group having six elements consecutive and increasing in atomic number. Each element is given an identifying letter (not its chemical symbol) and there is no relationship between the atomic numbers of one group and those of any other group.

Group A	Group B	Group C	Group D
a 1320	a 1530	a 587	a 727
b 2380	b 428	b 769	b 769
c 530	c 596	c 957	c 765
d 906	b 637	b 950	b 748
e 810	c 667	c 1150	c 755
f 1090	f 672	f 1360	f 915

(a) Which Group(s) contain alkali metals? (2 marks)
(b) Which Group(s) contain transition metals? (2 marks)
(c) Which Group(s) contain only transition metals? (1 mark)
(d) Plot out on a graph the ionization energies of the elements in Group A

against atomic number and use it to decide what type of electron (s, p, d etc.) is being removed when the elements c to f are ionized. (3 marks)

(e) What would you expect the ionization energy of the next element (g) in Group A to be? (2 marks)

SECTION 5

The Periodic Table 3: the elements of Group VII

5A Fixed response items

1 In which of the following compounds is sulphur most oxidized?

   A H_2S **B** Na_2SO_3 **C** $CuSO_4$ **D** $KHSO_3$ **E** NaHS

2 When sodium thiosulphate ($Na_2S_2O_3$) reacts with iodine in neutral solution

   A the iodine is oxidized to iodide ions (I^-)
   B the sulphur atoms in the thiosulphate ion are reduced to sulphur which is precipitated
   C the thiosulphate ion gains electrons to make the tetrathionate ion
   D the iodine is reduced to iodide ions (I^-)
   E the oxygen atoms in the thiosulphate ion are reduced to oxidation state $+2$

3-8 Each of these items concerns the following situation:
When concentrated hydrochloric acid is heated with solid potassium dichromate $K_2Cr_2O_7$, a pungent smelling gas is evolved which bleaches litmus. The other products of the reaction include chlorides of potassium and chromium, together with water.

3 The gas evolved is probably

   A chlorine   **B** oxygen   **C** chromium chloride
   D hydrogen chloride   **E** bromine

4 The oxidation number of chlorine in hydrochloric acid is

   A $+1$ **B** -1 **C** 0   **D** $+2$ **E** -2

5 The oxidation number of hydrogen in hydrochloric acid is

   A $+1$ **B** -1 **C** 0   **D** $+2$ **E** -2

6 The oxidation number of potassium in potassium dichromate is

   A $+1$ **B** -1 **C** 0   **D** $+2$ **E** -2

7 The oxidation number of the chromium in this reaction changes as follows:

- **A** it starts at +6 and gets more positive
- **B** it starts at −6 and gets less negative
- **C** it does not change
- **D** it starts at +6 and gets less positive
- **E** it starts at −6 and gets more negative

8 The oxidation number of the oxygen of the potassium dichromate changes as follows during this reaction:

- **A** it starts at +2 and gets more positive
- **B** it starts at −2 and gets less negative
- **C** it does not change
- **D** it starts at +2 and gets less positive
- **E** it starts at −2 and gets more negative

9–12 In each of these items, there is an unbalanced redox equation. Balance the equation, then select from the following lettered list the number in front of the first species in the equation:

A 2 **B** 3 **C** 4 **D** 5 **E** 6

9 $HI(aq) + H_2O_2(aq) \rightarrow H_2O(l) + I_2(aq)$

10 $Fe^{2+}(aq) + Cr_2O_7^{2-}(aq) + H^+(aq) \rightarrow Cr^{3+}(aq) + H_2O(l) + Fe^{3+}(aq)$

11 $MnO_4^-(aq) + Br^-(aq) + H^+(aq) \rightarrow Br_2(aq) + Mn^{2+}(aq) + H_2O(l)$

12 $NaOH(aq) + Cl_2(g) \rightarrow NaClO_3(aq) + NaCl(aq) + H_2O(l)$

13 When iodide ions react with iodate ions in acid solution the following reaction takes place:

$$5I^-(aq) + IO_3^-(aq) + 6H_3O^+(aq) \rightarrow 3I_2(aq) + 9H_2O(l)$$

In this reaction:

1. the oxidation number of the hydrogen changes from 0 to +1
2. the iodine in the iodate ion changes in oxidation number from −5 to 0
3. the iodate ion acts as an oxidizing agent

14 A white powder (Z) is heated and dry hydrogen halide gas (Y) is passed over it. After the reaction a white solid (X) remains which gives a very pale yellow precipitate when dissolved in water and treated with silver nitrate solution. When X is warmed with manganese(IV) oxide and concentrated sulphuric acid a red vapour is formed. From this it may be deduced that

1. X contained iodine
2. Z could have been the oxide of a metal
3. Y was hydrogen bromide

15 A set of silver halide precipitates is prepared and allowed to stand in sunlight. Which of the following statements is/are true?

1 the silver chloride darkens
2 the silver iodide is yellow
3 the silver iodide becomes dark grey

16 When silver nitrate solution is added to each of the following, in which case will there be no significant observation to be made?

A sodium iodide   B potassium bromide
C hydrochloric acid D sodium fluoride
E hydrobromic acid

17 When dilute ammonia is added to a set of silver halide precipitates, which of the following happens?

A both the silver bromide and the silver iodide dissolve
B both the silver chloride and the silver iodide dissolve
C silver chloride dissolves easily, silver iodide hardly at all
D silver iodide dissolves easily, the others hardly at all
E all the silver halides dissolve easily

18 When bromine reacts with sodium hydroxide solution

1 the colour of the bromine disappears
2 Br^- ions are formed
3 BrO^- ions are formed

19 The colour of iodine is apparently

1 purple when it is vapourized
2 brown when it is dissolved in potassium iodide solution
3 purple when it is dissolved in 1,1,1-trichloroethane

20 50 cm³ of a solution of bromine in water is treated with excess potassium iodide and the resulting iodine reacts exactly with 10 cm³ of 0.01 M sodium thiosulphate:

$$2S_2O_3^{2-}(aq) + I_2(aq) \rightarrow S_4O_6^{2-}(aq) + 2I^-(aq)$$

The concentration of the bromine solution in mol dm⁻³ is:

A 1×10^{-3} B 2×10^{-3} C 5×10^{-3}
D 1×10^{-2} E 2×10^{-2}

5B Structured questions

1 Copy the chart on the next page and insert on it the formulae which follow, putting each formula on the line corresponding to the oxidation number of the phosphorus atom(s) which it contains.

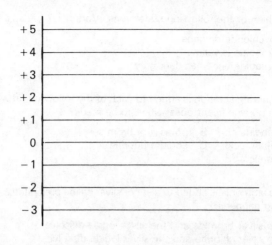

P$_4$O$_6$ P$_4$O$_{10}$ Na$_2$HPO$_4$ H$_2$PO$_4^-$
H$_3$PO$_3$ H$_3$PO$_2$ PCl$_3$ PCl$_5$ PH$_3$ P$_2$H$_4$
(1 mark each)

2 Here is an extract from a practical text book[5] describing the preparation of iodine(III) chloride, which is a volatile yellow solid:

'Set up a 500 cm³ round bottomed flask fitted with a gas inlet tube and a calcium chloride drying tube. Attach a supply of dry chlorine to the gas inlet tube and put into the flask 0.5 g of dry iodine crystals. Carry out the experiment in a fume cupboard.
Allow a slow stream of chlorine to enter the flask and sublime the iodine using a small flame. Yellow iodine(III) chloride is deposited on the cooler parts of the flask. When the iodine appears to have reacted, turn off the chlorine supply, cool the flask and remove the solid by scraping it out with a spatula.'

(a) What can be deduced from the insistence upon dryness throughout these instructions? (1 mark)
(b) What might you expect to see when you heated the iodine? (1 mark)
(c) How would you attempt to dry a sample of iodine for this experiment? (2 marks)
(d) Sketch the apparatus described in the first paragraph and indicate on it where you would expect the iodine(III) chloride to be deposited. (3 marks)
(e) How would you know when all the iodine had reacted? (1 mark)
(f) Write an equation for the reaction. (2 marks)

3 Iron(III) ions react with iodide ions to give iodine. 0.0163 g of iron(III) chloride is reacted with excess of potassium iodide solution. The iodide produced is titrated with 0.01 M sodium thiosulphate. 10 cm³ was required for complete reaction. (Fe = 56, Cl = 35.5)

(a) How many moles of iodine molecules were produced in the first reaction? (1 mark)
(b) How many moles of iodide ions must have reacted? (1 mark)
(c) How many moles of iron(III) ions reacted? (1 mark)
(d) Deduce the equation for the reaction between iron(III) ions and iodide ions. (2 marks)
(e) What indicator could be used for the titration and what is unusual about the way it is used? (2 marks)
(f) Mention one use of iodine or one of its compounds. (1 mark)
(g) By what process is iodine manufactured (mention only the raw material and an outline of the process)? (2 marks)

SECTION 6

Energy changes and bonding

6A Fixed response items

1 For which of the following changes can ΔH *not* be measured directly?

   A $He(g) - e^- \rightarrow He^+(g)$
   B $H_2SO_4(l) + 2H_2O(l) \rightarrow 2H_3O^+(aq) + SO_4^{2-}(aq)$
   C $Mg^{2+}(g) + 2Cl^-(g) \rightarrow MgCl_2(s)$
   D $NaOH(s) + aq \rightarrow Na^+(aq) + OH^-(aq)$
   E $CH_3COCH_3(l) + 4O_2(g) \rightarrow 3CO_2(g) + 3H_2O(l)$

2 In the energy level diagram shown, each energy change is accompanied by a lettered description of that energy change. Which of these descriptions is incorrect?

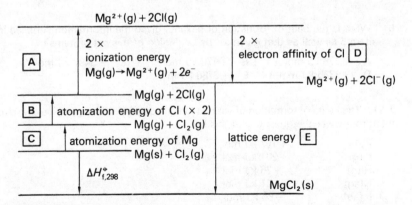

3 In a refinery process used to recover sulphur from petroleum, sulphur compounds are catalytically converted to hydrogen sulphide (H_2S) which is then mixed with a controlled amount of air and burned in a reaction furnace under carefully controlled conditions. The products of the reaction are $H_2O(g)$ and $S(s)$. What is the heat of reaction per mole of hydrogen sulphide if the heats of formation of $H_2S(g)$ and $H_2O(g)$ are -20.5 kJ mol^{-1} and -243.0 kJ mol^{-1} respectively?

A $+222.5$ kJ mol^{-1} **B** -222.5 kJ mol^{-1} **C** -445.0 kJ mol^{-1}
D $+202.0$ kJ mol^{-1} **E** -202.0 kJ mol^{-1}

4-6 These items concern the heats of combustion of the first seven alkanes which are as follows:

methane CH_4	-895 kJ mol^{-1}
ethane C_2H_6	-1562 kJ mol^{-1}
propane C_3H_8	-2230 kJ mol^{-1}
butane C_4H_{10}	-2925 kJ mol^{-1}
pentane C_5H_{12}	-3520 kJ mol^{-1}
hexane C_6H_{14}	-4175 kJ mol^{-1}
heptane C_7H_{16}	-4825 kJ mol^{-1}

4 The difference between successive values of ΔH (combustion) is a measure of

A the energy required to break two C—C bonds and two C—H bonds
B the energy required to break one C—C bond and three C—H bonds
C the energy required to break two C—C bonds and one C—H bond
D the energy required to break three C—C bonds and three C—H bonds
E the difference between the energy required to break some bonds and that produced in forming some others

5 If the heat of formation of $CO_2(g)$ is -395 kJ mol^{-1} and that of $H_2O(l)$ is -283 kJ mol^{-1}, what is the energy required to break a C—H bond?

A $+66$ kJ mol^{-1} **B** -217 kJ mol^{-1} **C** $+16.5$ kJ mol^{-1}
D -54.25 kJ mol^{-1} **E** it cannot be calculated unless one knows the heat of atomization of carbon and hydrogen

6 What is the heat of formation of hexane, given the information supplied in item **5** as well as that given at the beginning of this set of items?

A -176 kJ mol^{-1} **B** $+176$ kJ mol^{-1} **C** $+3497$ kJ mol^{-1}
D $+3786$ kJ mol^{-1} **E** -3786 kJ mol^{-1}

7-11 The heats of formation of various iodine compounds are given below. Use them to answer the questions in this set.

$I(g)$	$+107$ kJ mol^{-1}
$I^-(g)$	-201 kJ mol^{-1}
$HI(g)$	$+26$ kJ mol^{-1}
$I_2(aq)$	$+21$ kJ mol^{-1}
$I^-(aq)$	-56 kJ mol^{-1}

7   Which of the following represents the heat of atomization of iodine?

   A   +214 kJ mol⁻¹ (i.e. 2 × 107)   B   −214 kJ mol⁻¹
   C   +107 kJ mol⁻¹   D   −107 kJ mol⁻¹
   E   +53.5 kJ mol⁻¹ (i.e. ½ × 107)

8   Which of the following additional pieces of information would you require to calculate the heat of the reaction:

   $H_2(g) + I_2(g) \rightarrow 2HI(g)$

   A   None, we already have the answer as 26 kJ mol⁻¹
   B   None, the heat of formation of an element is always zero
   C   The heat of atomization of iodine
   D   The heat of formation of gaseous iodine
   E   The heat of formation of gaseous hydrogen

9   If the heat of formation of gaseous potassium ions is +510 kJ mol⁻¹ the lattice energy of potassium iodide is

   A   −309 kJ mol⁻¹   B   −711 kJ mol⁻¹
       (i.e. +201 − 510)   (i.e. −201 − 510)
   C   +309 kJ mol⁻¹   D   +711 kJ mol⁻¹
   E   none of these, we have insufficient information to find it

10 When hydrogen iodide dissolves in water, the following change takes place:

   $HI(g) + H_2O(l) \rightarrow H_3O^+(aq) + I^-(aq)$

   (this is formally an equilibrium, but it may be taken to go to completion). Which of the following additional pieces of information would you require to calculate the heat of this reaction?

   A   The heat of formation of $H_3O^+(aq)$
   B   The heat of formation of $H_2O(l)$
   C   The heat of formation of $H_3O^+(aq)$ and of $HI(g)$
   D   The heat of formation of $H_3O^+(aq)$ and of $H_2O(l)$
   E   The heat of the reaction $H_2O(l) + H^+(g) \rightarrow H_3O^+(aq)$

11 Which of the following named energy quantities is given by $\Delta H_f^\ominus(I^-(g)) - \Delta H_f^\ominus(I(g))$ i.e. −201 − 107 kJ mol⁻¹?

   A   the heat of solution of iodine
   B   the first ionization energy of iodine
   C   the atomization energy of iodine
   D   the electron affinity of iodine
   E   the heat of formation of gaseous iodine

12 Some bond energy terms $\overline{E}$ in kJ mol⁻¹ are

   $\overline{E}_{H-H} = 436$ $\overline{E}_{Cl-Cl} = 242$ $\overline{E}_{H-Cl} = 431$

   The heat of formation of gaseous HCl in kJ mol⁻¹ is

   A   −184   B   −92   C   −368   D   −625   E   −237

13-16 These items concern the Born–Haber cycle:

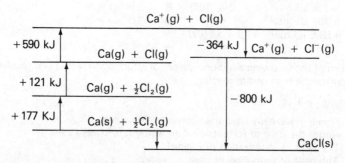

13 The heat of formation of solid calcium(I) chloride in kJ mol^{-1} is

    A -800 **B** -364 **C** $+800$ **D** $+276$ **E** -276

14 Which of the following energies shown on the diagram is likely to be the least reliable in accuracy?

    A $+177$ **B** $+121$ **C** $+590$ **D** -364 **E** -800

15 Which of the following statements about CaCl(s) is true?

    A it contains chlorine in an unusual oxidation state
    B it is stable with respect to its elements
    C it contains calcium in its most usual oxidation state
    D it is a covalent compound
    E it is the usual chloride of calcium

16 Which of the following is most likely to happen to CaCl(s) on standing?

    A it would change into calcium(I) oxide
    B it would disproportionate
    C it would absorb water from the air
    D it would hydrolyse
    E it would decompose into its elements

17-20 These items concern entropy and enthalpy changes involving the reaction whose equation is:

$$CH_3OH(l) + HBr(g) \rightarrow CH_3Br(g) + H_2O(l) \quad \Delta H^{\ominus} = -46.0 \text{ kJ mol}^{-1}$$

Data: $S^{\ominus}$ /J mol^{-1} K^{-1}
$CH_3OH(l)$ 127.2
$HBr(g)$ 198.5
$CH_3Br(g)$ 246.2
$H_2O(l)$ 70.0

17 What is the entropy change $\Delta S^{\ominus}_{\text{system}}$ in J mol^{-1} K^{-1}?

    A $+9.5$ **B** -9.5 **C** 0 **D** $+154$ **E** -154

18 What is the entropy change $\Delta S^{\ominus}_{\text{surroundings}}$ in J mol^{-1} K^{-1}?

    A $+9.5$ **B** -9.5 **C** 0 **D** $+154$ **E** -154

19 What is true of the entropy changes for a spontaneous chemical process?

   A the entropy of the system must increase
   B the entropy of the surroundings must increase
   C the entropy of the surroundings must decrease
   D the total entropy of the system and surroundings must increase
   E the total entropy of the system and surroundings must decrease

20 If, in another reaction similar to the one mentioned,

$\Delta H^\ominus = -20$ kJ mol^{-1}

$\Delta S^\ominus = -8$ J mol^{-1} K^{-1}

what would be the total standard entropy change for system and surroundings in J mol^{-1} K^{-1} at 298 K?

   A -28 **B** $-20\,008$ **C** -8 **D** $+59$ **E** -59

6B Structured questions

1 In an experiment[6] to estimate the bond energy term for the bond C—Br the heat of the reaction between cyclohexene and bromine was determined.

cyclohexene   1,2-dibromocyclohexane

50 cm³ of a solution containing 10 g of bromine per dm³ of tetrachloromethane was placed in an insulated flask fitted with a stirrer, an electrical heating coil and a sensitive thermometer. A small excess of cyclohexene was added and, after stirring, the maximum temperature rise was found to be 3.2 K. The thermal capacity of the flask, heating coil, thermometer and liquids was estimated electrically and found to be 0.127 kJ per degree.

(a) What mass of bromine was used in the experiment? (1 mark)
(b) How much heat was produced from this mass of bromine? (1 mark)
(c) How much heat would have been produced from 1 mole of bromine molecules Br_2? This is ΔH for the reaction. (1 mark)
(d) When cyclohexene reacts with bromine, what bonds are broken? (2 marks)
(e) When cyclohexene reacts with bromine, what bonds are formed? (1 mark)
(f) Draw an energy level diagram for this reaction. (2 marks)
(g) Calculate a value for the bond energy term of the C—Br bond. (2 marks)

Data: bond energy terms C=C   613 kJ mol^{-1} relative atomic mass
                                 C—C   347 kJ mol^{-1} of bromine = 80
                                 Br—Br 197 kJ mol^{-1}

2 It has been reported that the anaerobic sulphate-reducing bacterium *Desulphovibrio desulphuricans* obtains its energy from the reduction of sulphate ions in the presence of a reductant. The reaction does not take place in the absence of the bacteria. The equation for the reaction is:

$$SO_4^{2-}(aq) + 4H_2(g) \rightarrow S^{2-}(aq) + 4H_2O(l)$$

Heats of formation $\Delta H^\ominus_{f,298}$ are as follows:

$SO_4^{2-}(aq)$ -910 kJ mol^{-1}
$S^{2-}(aq)$ $+32.7$ kJ mol^{-1}
$H_2O(l)$ -286.2 kJ mol^{-1}

(a) If the sulphate ions were present as dilute sulphuric acid, and given that $S^{2-}(aq)$ is a strong base, in what form would you expect the sulphur to be at the end of the reaction? (2 marks)
(b) Calculate the heat of the reaction as shown in the equation. (3 marks)
(c) Is the reaction exothermic or endothermic? (1 mark)
(d) Does your answer to the previous question suggest that the reaction should be possible? (2 marks)
(e) Why do you think the reaction 'does not take place' in the absence of the bacteria? What difference does the presence of the bacteria make? (2 marks)

3 This question concerns the following energy data about hydrogen and chlorine.

Heat of atomization of hydrogen = 219 kJ mol^{-1}
Heat of atomization of chlorine = 123 kJ mol^{-1}
Electron affinity of hydrogen = 69 kJ mol^{-1}
Electron affinity of chlorine = -365 kJ mol^{-1}

The chloride ion is Cl^-, the hydride ion is H^-.

(a) What is the heat of formation of (i) hydride ions, (ii) chloride ions? (4 marks)
(b) Which are the more stable with respect to the element, hydride ions or chloride ions? (2 marks)
(c) Some metals form ionic hydrides containing the ion H^-. Some metals form ionic chlorides containing the ion Cl^-. What further energy data would you require in order to compare the heats of formation of the hydride and the chloride of a given metal? (2 marks)
(d) A text book description of ionic hydrides mentions that they are more likely to be formed by metals with low ionization energies than by those with high ionization energies. Comment on the validity of this statement by considering what other energy factors ought to be taken into account. (2 marks)

SECTION 7

Crystal structure

7A Fixed response items

1. Which of the following best represents a metal lattice with body-centred cubic packing?

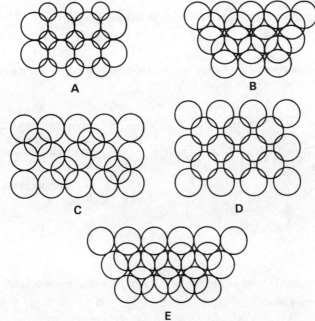

2. Which of the following substances would you expect to be anisotropic?

   A sodium iodide NaI
   B calcium carbonate $CaCO_3$
   C diamond C
   D sodium sulphide Na_2S
   E copper chloride $CuCl_2$

3-5 The zinc blende structure (ZnS) may be represented thus:

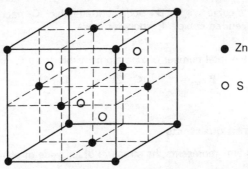

3 What fraction of a zinc atom is at each corner of the unit cell?

    A $\frac{1}{2}$ **B** $\frac{1}{4}$ **C** $\frac{1}{6}$ **D** $\frac{1}{8}$ **E** $\frac{1}{16}$

4 What is the arrangement of the zinc atoms?

    A face-centred cubic   **B** simple cubic
    C body-centred cubic   **D** tetrahedral
    E hexagonal

5 What is the coordination number of the sulphur?

    A 3   **B** 4   **C** 6   **D** 8   **E** 12

6-10 The questions which follow concern the following diagram of the unit cell of Wurtzite, another form of zinc sulphide:

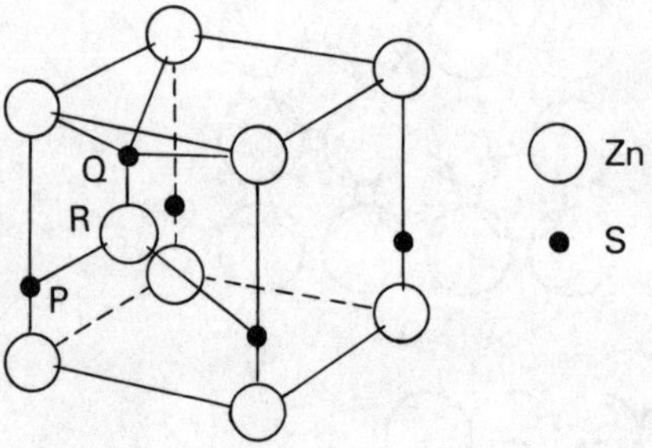

6 Consider the sulphur atom P. What fraction of this atom is within the unit cell?

    A a whole atom
    B $\frac{1}{2}$
    C $\frac{1}{4}$
    D $\frac{1}{6}$
    E $\frac{1}{8}$

7 Consider atom Q. What fraction of this atom is within the unit cell?

    A a whole atom   **B** $\frac{1}{2}$ **C** $\frac{1}{4}$ **D** $\frac{1}{6}$ **E** $\frac{1}{8}$

8 Consider the zinc atom R. What fraction of this atom is completely within the unit cell?

    A a whole atom   **B** $\frac{1}{2}$ **C** $\frac{1}{4}$ **D** $\frac{1}{6}$ **E** $\frac{1}{8}$

9 How would you describe the arrangements of zinc atoms in this structure?

    A simple cubic   **B** body-centred cubic   **C** face-centred cubic
    D side-centred cubic   **E** none of these

10 What is the total number of zinc atoms within the unit cell?

    A 1   **B** 2   **C** $2\frac{1}{2}$ **D** $2\frac{3}{4}$ **E** 3

7B Structured questions

1 The diagram[7] represents the structure of an oxide of platinium.

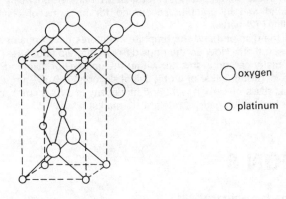

(a) What is the co-ordination number of the oxygen atoms in this structure? (1 mark)
(b) What is the co-ordination number of the platinum atoms in this structure? (1 mark)
(c) What is the formula for the platinum oxide? (1 mark)
(d) How are the nearest oxygen atoms to each platinum atom arranged around that atom? (1 mark)
(e) How are the nearest platinum atoms to each oxygen atom arranged around that atom? (1 mark)
(f) Would it be reasonable to describe the region of the structure bounded by the dotted lines as a 'unit cell' of this structure? Justify your answer. (2 marks)
(g) This structure may be regarded as a cube of oxygen atoms with platinum atoms at the centres of two opposite faces, above a similar cube turned through 90° about a vertical axis. Using the diagram above as a guide, draw the structure in such a way as to emphasize this view of it. (3 marks)

2

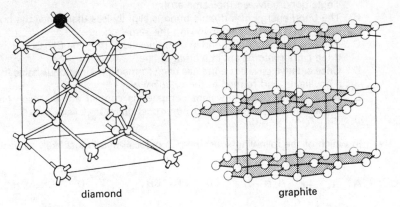

diamond　　　　　　　　　　　　graphite

(a) What is the co-ordination number of carbon in diamond? (1 mark)
(b) What is the co-ordination number of carbon in each plane of graphite? (1 mark)

(c) How is the difference in co-ordination number between carbon in diamond and in graphite accounted for in terms of electrons and bonding? (2 marks)
(d) Both the diamond and the graphite structures have rings of six carbon atoms in them. How do the rings differ in the two structures? (2 marks)
(e) How many carbon atoms are within the unit cell of diamond? (1 mark)
(f) If the length of 1 side of a unit cell of diamond is 0.357 nm, the relative atomic mass of carbon is 12 and its density is 3.53 g cm^{-3}, calculate a value for the Avogadro Constant. (3 marks)

SECTION 8

Chemical bonding

8A Fixed response items

1 In which of the following ionic compounds would there be a significant discrepancy between calculated and experimental lattice energies?

1 ZnS   **2** AgBr   **3** NaF

2 Correct statements about the bonding in the chlorate(V) ion ClO_3^- include

1 the chlorine atom has a lone pair of electrons
2 there is delocalization of electrons in the ion
3 all the O—Cl—O bond angles are identical

3 Which of the following statements about various types of chemical bond is UNtrue?

A A double covalent bond between two atoms is generally shorter than a single bond between the same atoms
B The bond energy of a double bond is slightly less than twice the bond energy of a single bond involving the same atoms
C Because of the double bond in the molecule of propene CH_2CHCH_3 the three carbon atoms are in a straight line
D Once a dative covalent bond has been formed it is indistinguishable from a covalent bond formed in the ordinary way
E A number of stable covalent compounds contain atoms which have more than 8 electrons in their outermost quantum levels

4 In which of the following substances is delocalization most likely to occur?

A (benzene ring with O—H substituent)

B $CH_3\overset{O}{\overset{\|}{C}}CH_3$

C $CH_3CH_2\overset{}{\underset{\|}{C}}CH_2OH$ with $\|$ O below

D $CH_3\overset{O}{\overset{\|}{C}}OH$

E $NH_2\overset{}{\underset{\|}{C}}NH_2$ with $\|$ O below

5 In the oxalate ion, for which one possible structural formula is

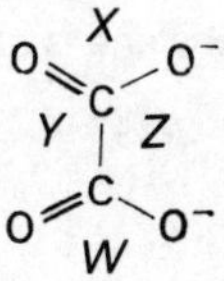

$$\begin{array}{c} O \underset{Y}{\overset{X}{=}} C \underset{W}{\overset{Z}{-}} O^- \\ O = C - O^- \end{array}$$

1 there are two sets of carbon-to-oxygen bond lengths
2 the angles X, Y and Z are all identical
3 the angle X is equal to the angle W

6 In hydrogen peroxide H_2O_2

1 each oxygen atom has two lone pairs of electrons
2 the two hydrogen atoms are bonded to the same oxygen atom
3 the molecule is linear

7-10 Choose from the following lettered list the angle which most closely corresponds to the bond angle mentioned in each item.

A 109.5°
B slightly less than 109.5°
C slightly more than 109.5°
D 120°
E slightly less than 120°

7 The angle H—C—H in ethene C_2H_4

8 The angle C—N—C in trimethylamine $(CH_3)_3N$

9 The angle O=S=O in sulphur trioxide SO_3

10 The angle O⋯Si⋯O in the silicate ion $SiO_3{}^{2-}$

8B Structured questions

1 This question principally concerns the bond diagrams shown below:

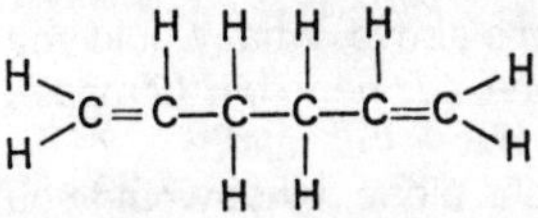

$$\begin{array}{c} H \\ H \end{array} C=C-C-C-C=C \begin{array}{c} H \\ H \end{array}$$ (with H's on middle carbons)

1,5-hexadiene

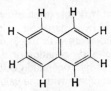
naphthalene

cyclohexanone

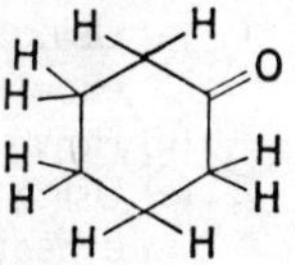

(a) How many lone pairs of electrons are there in the outer quantum level of the oxygen atom in cyclohexanone? (1 mark)
(b) Would the angle between the bonds connecting the carbon atoms in cyclohexanone be greater than, less than or the same as the angle between the bonds connecting the carbon atoms in naphthalene? Justify your answer. (2 marks)
(c) Draw a 'dot and cross' diagram to represent the distribution of electrons in this part of the molecule of 1,5-hexadiene: (2 marks)

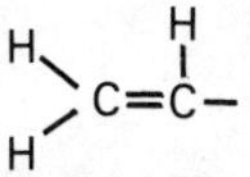

$$\begin{array}{c} H \\ H \end{array} C=C \begin{array}{c} H \\ - \end{array}$$

(d) Redraw the cyclohexanone structure and show on it the polarity of any polarized bonds. (1 mark)

(e) The delocalization of electrons in the molecule of naphthalene is sometimes represented by this diagram:

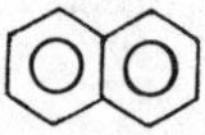

by analogy with the structure of benzene. Suggest why some people criticize this way of representing naphthalene. (2 marks)

(f) Are all the atoms of the naphthalene structure in the same plane? Justify your answer. (2 marks)

2   Three compounds, cyclohexene, 1,3-cyclohexadiene and benzene, can all be converted by reaction with 1, 2 and 3 molecules of hydrogen respectively into cyclohexane, in reactions of a type called 'hydrogenation'. Simple bond structures for these compounds are as follows:

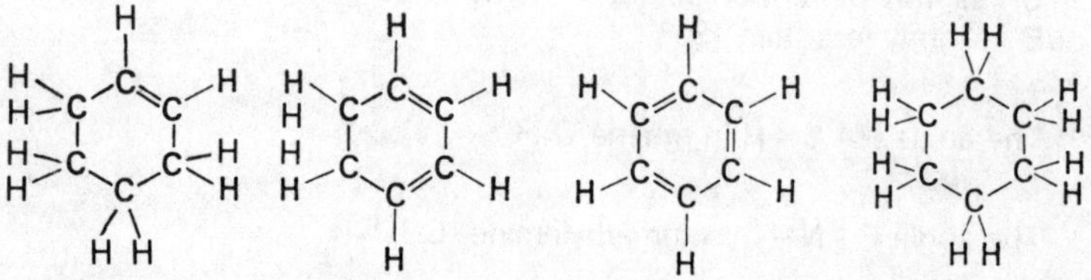

cyclohexene   **1,3-cyclohexadiene**   **benzene**   **cyclohexane**

The heats of hydrogenation, i.e. the heat evolved when each of the first three compounds is converted into cyclohexane, are as follows:

cyclohexene	-121 kJ mol^{-1}
1,3-cyclohexadiene	-233 kJ mol^{-1}
benzene	-209 kJ mol^{-1}

(a) Using the heat of hydrogenation of cyclohexene above, what would you expect the heat of hydrogenation of 1,3-cyclohexadiene to be? (2 marks)
(b) How does this compare with the value given above? (1 mark)
(c) Using the heat of hydrogenation of cyclohexene alone, what would you expect the heat of hydrogenation of benzene to be? (2 marks)
(d) How does this compare with the value given above? (1 mark)
(e) What phenomenon accounts for the difference between the answers to (b) and (d)? (2 marks)
(f) Is it possible to draw other bond arrangements than the one shown above for 1,3-cyclohexadiene? If so, draw one of them. (2 marks)

SECTION 9

Organic chemistry 1: hydrocarbons and halogenoalkanes

9A Fixed response items—hydrocarbons

1   Which of the following would react the most readily with bromine solution?

    A $CH_3CH_2CH_3$ **B** CH_2CHCH_3 **C** $CH_3CH(CH_3)CH_3$

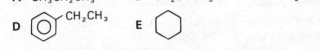

2   Which of the following would you *not* expect to find amongst the products when ethene is shaken with water containing bromine, chloride ions and iodide ions?

    A CH_2ClCH_2Cl **B** CH_2ClCH_2Br **C** CH_2BrCH_2Br
    D CH_2BrCH_2I **E** CH_2OHCH_2Br

3   Which of the following is isomeric with pentan-1-ol?

    A $CH_3CH_2CH(OH)CH_3$ **B** $CH_3CH_2COCH_2CH_3$
    C $CH_3OCH(CH_3)CH_2CH_3$ **D** $CH_3OCH_2CH_2CH_2CH_2OH$
    E $CH_3CH_2CH_2CH_2CO_2H$

4   Which of the following would you expect to react vigorously and exothermically with concentrated sulphuric acid?

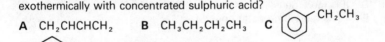

5   Which of the following molecular formulae could correspond to an alkene with two double bonds?

    A C_4H_{10} **B** C_7H_9 **C** C_5H_{12} **D** C_6H_{10} **E** C_8H_{16}

6   When organic compounds burn with smoky flames, which of the following is UNlikely to be found amongst the several combustion products?

    A water   **B** carbon monoxide   **C** carbon dioxide
    D carbon   **E** hydrogen

7-10 In these items, classify the substance described into one of the following lettered types:

    A alkene   **B** alkane   **C** arene
    D cyclic alkane   **E** cyclic alkene

7   A compound of formula C_7H_8 which burns very smokily and reacts only very slightly with acidified potassium manganate(VII)

8 A compound of formula C_3H_8 which does not react with a solution of bromine in tetrachloromethane in the dark

9 A compound of formula C_6H_{12} which does not react with concentrated sulphuric acid

10 A compound which reacts with steam in the presence of phosphoric acid to give $CH_3CH(OH)CH_3$

11 Which of the following is the product when but-2-ene reacts with HBr?

    A $CH_3CH_2CH_2CH_2Br$ **B** $CHBrCHCH_2CH_3$ **C** $CH_3CBrCHCH_3$
    D $CH_3CHBrCHBrCH_3$ **E** $CH_3CHBrCH_2CH_3$

12 X is an organic liquid which is immiscible with water. When it is shaken with bromine water, the colour of the bromine is transferred to the X layer. Which of the following could X be?

    A   ethane C_2H_6 **B**   ethene C_2H_4 **C**   benzene C_6H_6
    D   ethanol C_2H_5OH **E**   hexene C_6H_{12}

13 Phosphoric acid is used in preference to sulphuric acid for dehydrating alcohols because

    A   it is a relatively poor oxidizing agent
    B   it is the more volatile
    C   it is more miscible with water
    D   it is more dense
    E   it has more protons to donate

14 When benzene is attacked by a mixture of nitric and sulphuric acids, which of the following functions does nitric acid have in the mechanism?

    A   oxidizing agent   **B**   reducing agent   **C**   base
    D   solvent   **E**   dehydrating agent

15 Which of the following conditions would it be necessary to use in order to make benzene react with chlorine to give C_6H_5Cl?

    A   ultraviolet light   **B**   a halogen carrier
    C   aluminium oxide catalyst   **D**   platinum catalyst
    E   nickel catalyst

16 0.25 moles of a certain alkene react with 11.2 dm³ of hydrogen (measured at s.t.p.). How many double bonds are there in each molecule of the alkene?

    A   1   **B**   2   **C**   3   **D**   4
    E   it depends on whether the carbon atoms are in a chain or a ring

17 Which of the following results from a reaction between benzene and concentrated sulphuric acid?

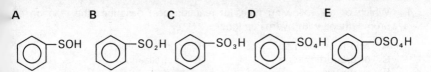

18 Which of the following results from the reaction of benzene with a mixture of nitric and sulphuric acids?

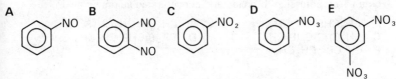

19 If ethene reacts with chlorine in the presence of fluoride, chloride, bromide and iodide ions and the resulting mixture is subjected to gas chromatography, which of the following would emerge from the chromatographic column first?

A $ClCH_2CH_3$
B $ClCH_2CH_2F$
C $ClCH_2CH_2Cl$
D $ClCH_2CH_2Br$
E $ClCH_2CH_2I$

20 When bromine attacks the double bond in propene, which of the following is the formula of the ion formed after the first stage of the attack?

A $CH_3\overset{+}{C}HCH_2Br$
B $CH_3^-CHCH_2Br$
C $CH_3CBr\overset{+}{C}H_2$
D $CH_3CHBr\overset{-}{C}H_2$
E $CH_3\overset{+}{C}HCHBr$

9B Fixed response items—halogenoalkanes

1 If bromoethane is to be reacted with potassium cyanide, which of the following solvents is used?

A water
B ether
C ethanol
D tetrachloromethane
E hexane

2 Two drops of each of the following are mixed with ethanol and aqueous silver nitrate solution. Which halide would give a precipitate first?

A chlorobenzene
B 2-chlorobutane
C 2-chloro 2-methyl propane
D 1-chlorobutane
E 1-fluorobutane

3 When bromobutane is reacted with excess ammonia which of the following is the formula for the main product?

A $CH_3CH_2CH_2CH_2NH_3$
B $CH_3CH_2CH_2CH_2NH_2$
C $(CH_3CH_2CH_2CH_2)_2NH$
D $(CH_3CH_2CH_2CH_2)_3N$
E $(CH_3CH_2CH_2CH_2)_3N^+Br^-$

4   Which of the following types of reaction best describes the reaction of bromoethane with hydroxide ions?

   A acid-base reaction   B redox reaction
   C disproportionation   D neutralization
   E hydrolysis

5   Which of the following would be the main product of the reaction between butan-1-ol, potassium bromide and concentrated sulphuric acid?

   A $CH_3CHBrCH_2CH_3$ B $CH_3CH_2CH_2CH_2Br$
   C $CH_3CHBrCHBrCH_3$ D $CH_2BrCHBrCH_2CH_3$
   E $CH_3CBr_2CH_2CH_3$

6   Which of the following is structurally different from the others?

   A CH_3Br B $CH_2BrCH_2CH_3$
   C CH_3CH_2Br D $CH_3CH_2CHBrCH_3$
   E $CH_3CH_2CH_2CH_2Br$

7   Which of the following is the formula for the main organic substance resulting from the reaction of 2-bromopropane with potassium hydroxide dissolved in ethanol?

   A CH_3CHCH_2 B $CH_3CH_2CH_3$ C $CH_3CH_2CH_2OH$
   D $CH_3CH(OH)CH_3$ E $CH_3CHBrCH_2OH$

8   Which of the following is the name corresponding to the formula $CH_3CH(CH_3)CHCH_2$?

   A pent-1-ene   B 3-methylbut-1-ene
   C 2-methylbut-3-ene   D 2-methylbut-4-ene
   E 2-methylpent-1-ene

9   Which of the following would not be a liquid at room temperature and pressure?

   A chloromethane   B iodoethane   C chlorobenzene
   D chloropentane   E bromopropane

10 In the preparation of a bromoalkane from an alcohol by reaction with a mixture of potassium bromide and conc. sulphuric acid, which of the following is *unlikely* to be amongst the reaction products?

   A hydrogen bromide   B bromine
   C potassium hydrogensulphate   D water
   E hydrogen

9C Structured question

7   You are provided with a copy of the instructions for a method[8] of preparing cyclohexene.

Small-scale preparation of cyclohexene

To 10 cm³ of cyclohexanol in a flask, add, with a dropping pipette, 4 cm³ of concentrated phosphoric acid, shaking the flask.

Assemble the apparatus (see below) and heat the flask gently, distilling over the liquid.

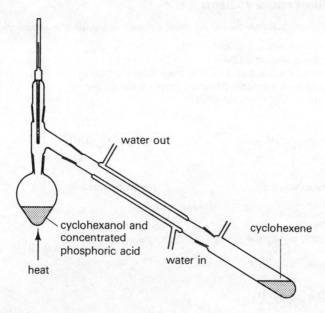

Pour the distillate into a separating funnel and add 2 cm³ of a saturated solution of sodium chloride. Shake the mixture and allow the two layers to separate. Run off the lower layer and then run the top layer, containing cyclohexene, into a small flask. Add 2 or 3 pieces of anhydrous calcium chloride, stopper the flask and shake until the liquid is clear.

Decant the liquid into a clean distillation flask and distil it, collecting the liquid boiling at 81–85°C.

(a) Write the formula for cyclohexanol. (2 marks)
(b) Why is there a vent tube (side arm) just after the condenser? (1 mark)
(c) Why should you not distil the mixture until there is nothing left in the flask? (1 mark)
(d) What is the purpose of the sodium chloride solution? (1 mark)
(e) What do you deduce from the fact that the cyclohexene is the *upper* layer? (1 mark)
(f) What is the calcium chloride for? (1 mark)
(g) What change of appearance in the cyclohexene would tell you that enough calcium chloride had been used? (2 marks)
(h) What would you conclude if the boiling point during the final distillation were 75° instead of 81–85°C? (1 mark)
(i) Write the equation for the complete combustion of cyclohexene C_6H_{10}. (2 marks)
(j) What is produced when cyclohexene reacts with hydrogen, and what conditions are used? (3 marks)

39

SECTION 10

Intermolecular forces

10A Fixed response items

1-3 In each question decide whether the pair of liquids mentioned would

  A be immiscible or only partly miscible
  B obey Raoult's Law
  C show a positive deviation from Raoult's Law
  D show a negative deviation from Raoult's Law
  E react together chemically

1 cyclohexanol and methylbenzene

2 CH_2Cl_2 dichloromethane and $CH_3CH_2CCH_3$ (=O) butanone

3 trichloroethene ($Cl_2C=CHCl$) and H_2O water

4 If each of the following pairs of substances are mixed, in which case would you expect the temperature to rise?

  A propane and tetrachloromethane B hydrogen chloride and propanone
  C ethanol and water   D propanone and cyclohexane
  E trichloromethane and water

5 Which of the following pairs of substances would you *not* expect to form hydrogen bonds with each other?

  A CH_2Cl_2 and $CH_3CH_2COCH_3$ B CH_3OH and CH_3CH_2OH
  C F^- and HF   D CH_3CH_2SH and CH_3SH
  E CH_3CO_2H and H_2O

6 In which of the following substances would you expect the forces of attraction between molecules to be *principally* hydrogen bonds?

  1 liquid hydrogen bromide HBr
  2 liquid hydrogen peroxide H—O—O—H
  3 liquid hydrogen fluoride HF

7-10 These items refer to the following information:

It has been established that crystalline boric acid (B(OH)$_3$) is hydrogen bonded into sheets of atoms with the following structure. Only a small part of one such sheet is illustrated.

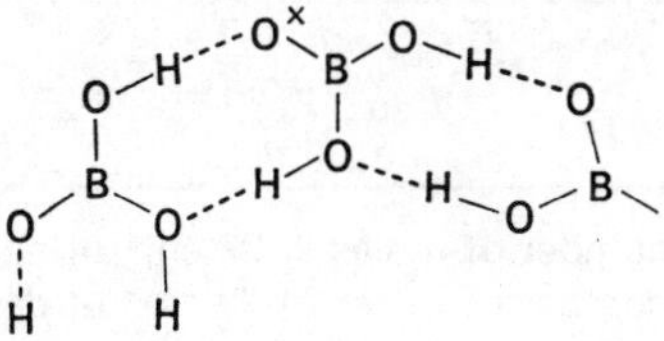

7. Which of the following most nearly represents the bond angle O—H---O?
   A 90°   B 109°   C 170°   D 180°   E 200°

8. Which of the following most nearly represents the bond angle O—B—O?
   A 90°   B 109°   C 120°   D 130°   E 170°

9. Consider the oxygen atom marked x. This oxygen atom is presumably joined to one other hydrogen atom. If the structure were continued so as to include this hydrogen atom, which of the following would best represent the situation?

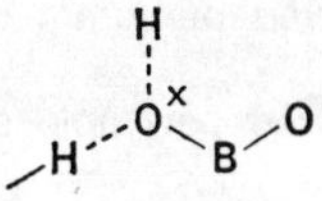

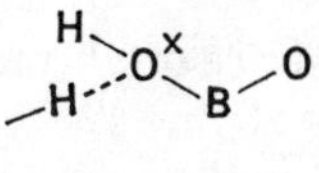

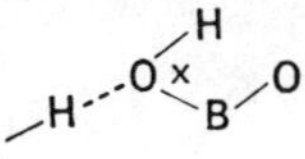

10. Within which of the following ranges would you expect the bond energy of the hydrogen bond in this structure to be?
    A 0.8–2.0 kJ mol^{-1} B 8.0–30.0 kJ mol^{-1} C 80–120 kJ mol^{-1}
    D 300–500 kJ mol^{-1} E 600–1000 kJ mol^{-1}

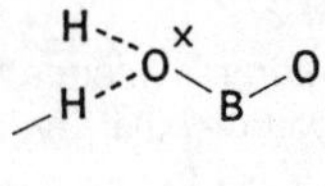

10B Structured questions

1. This question is about a mixture of 30 cm³ of hexane and 70 cm³ of decane. Some data are given in the table. Use the information to answer the questions.

	hexane	decane
density /g cm^{-3}	0.66	0.73
molar mass /g mol^{-1}	86	142
boiling point /K	342	447

   (a) (i) Calculate the number of moles in 30 cm³ of hexane.
     (ii) Calculate the number of moles in 70 cm³ of decane. (3 marks)
   (b) Calculate the fraction $\frac{\text{moles of decane}}{\text{total number of moles}}$ (1 mark)
   (c) What is the name given to this fraction? (1 mark)
   (d) Calculate the boiling point of the mixture. (1 mark)
   (e) What assumption did you make in calculating the boiling point of the mixture and how justified is it? (1 mark)
   (f) Account for the difference between the boiling points of pure hexane and pure decane. (1 mark)
   (g) Sketch a graph of the variation of vapour pressure of mixtures of hexane and decane with different compositions. (2 marks)

2. Mixing 50 g of propanone (RMM = 58) and 2 g of trichloromethane (RMM = 119) in a calorimeter of negligible heat capacity gave a temperature rise of 4 K. Specific heat capacities are:

   propanone   2.22 J g^{-1} K^{-1}
   trichloromethane 0.96 J g^{-1} K^{-1}

   (a) Calculate the heat evolved when the two liquids are mixed. (3 marks)
   (b) What would be the energy evolved per gram of trichloromethane? (3 marks)
   (c) What would be the energy evolved per mole of trichloromethane? (3 marks)
   (d) What is your answer to part (c) a measure of? (1 mark)

3. There is some evidence that in the molecule of 2-hydroxybenzoic acid there is an intramolecular hydrogen bond:

   (a) What does the word 'intramolecular' mean? (2 marks)
   (b) Consider the two atoms marked*. Are these co-planar with the benzene ring? Briefly justify your answer. (3 marks)

(c) By considering the bonds around the carbon atom marked* next and then the hydrogen atom attached to the oxygen atom marked*, decide whether the right-hand ring of atoms is co-planar. Briefly justify your answer. (3 marks)

(d) There is one unusual feature of the hydrogen bond in this molecule apart from anything which has been said already. What is this feature? (2 marks)

SECTION 11

Organic chemistry 2: —OH group compounds

11A Fixed response items

1 Which of the following is isomeric with propan-1-ol?

   A $CH_3CH_2OCH_2CH_3$ B $CH_3OCH_2CH_3$
   C CH_3COCH_3 D CH_3CH_2CHO
   E $CH_3CH_2COCH_3$

2 Which of the following is a secondary alcohol?

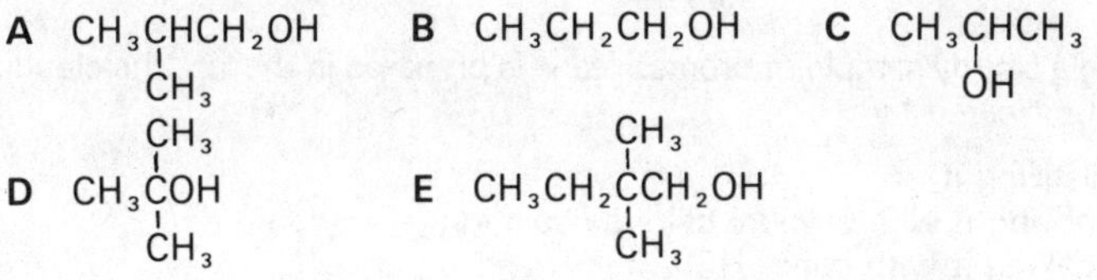

   A CH_3CHCH_2OH B $CH_3CH_2CH_2OH$ C CH_3CHCH_3
          $|$ $|$
         CH_3 OH

         CH_3 CH_3
         $|$ $|$
   D CH_3COH E $CH_3CH_2CCH_2OH$
           $|$ $|$
          CH_3 CH_3

3 The most satisfactory systematic name for the compound

      $CH_3CH(OH)CH(CH_3)CH_3$ is

   A 3-methylbutan-2-ol   B 3-methylbutyl alcohol
   C 2 hydroxypentane   D 1,2-dimethylpropan-1-ol
   E 2-methylhexan-1-ol

4 Phenol is a stronger acid than ethanol because

   1 it is more soluble in water than ethanol, so that phenol molecules have a better chance of donating protons to water molecules
   2 the ethanol molecule is stabilized by delocalization which inhibits the donation of a proton by the —OH group
   3 the phenoxide ion is stabilized by delocalization involving the benzene ring; this cannot happen with the ethoxide ion

5 Which of the following reaction mixtures will give an organic product which is ionic?

   1 C_2H_5OH, KBr, H_2SO_4 2 C_2H_5OH, PCl_5
   3 C_2H_5OH, Na

6 Which of the lettered atoms in the following formula changes in oxidation number when propan-1-ol is oxidized?

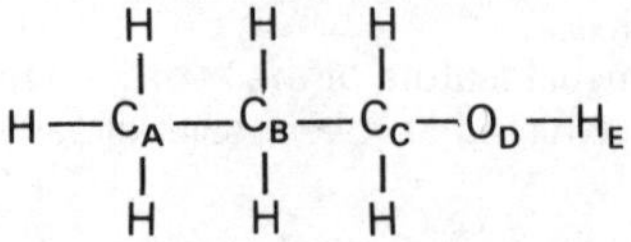

$$H-\underset{H}{\overset{H}{C_A}}-\underset{H}{\overset{H}{C_B}}-\underset{H}{\overset{H}{C_C}}-O_D-H_E$$

7 Which of the following has the highest boiling point?
A C_2H_5OH
B CH_3CHO
C CH_3CO_2H
D CH_3OH
E H_2O

8 When phenol reacts with sodium hydroxide, which of the following is the organic product?

A
B
C
D
E

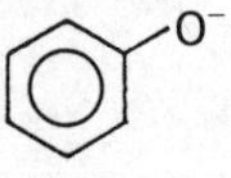

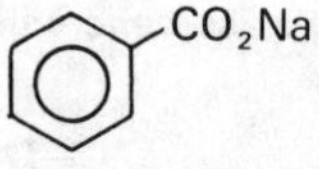

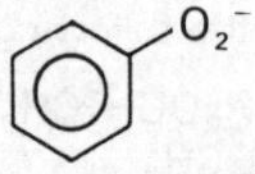

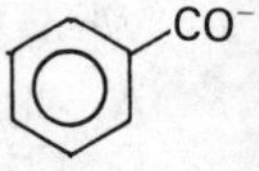

9 Which of the following results from the oxidation of propan-2-ol?
A C_2H_5OH
B CH_3CHO
C CH_3COCH_3
D C_2H_5CHO
E C_3H_7OH

10 When a cloudy sample of bromobutane is prepared in the lab, the cloudiness may be cleared by

A distilling it
B shaking it with calcium chloride solution
C shaking it with conc. HCl
D shaking it with sodium hydrogencarbonate
E drying it

11 Which of the following is the main product when phenol reacts with bromine water?

A
B
C
D, E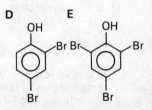

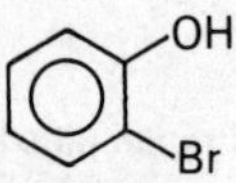

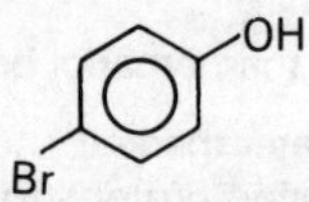

12 Which of the following would you expect to give a purple colour when mixed with neutral iron(III) chloride solution?

A, B, 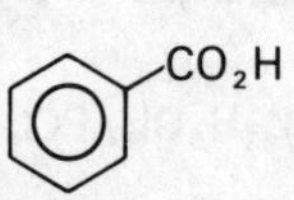 C, 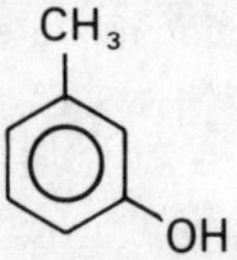 D, 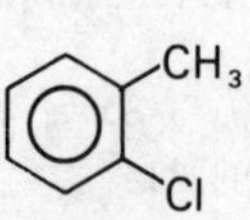 E

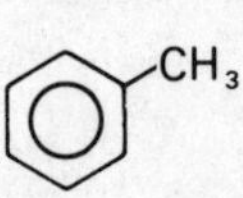

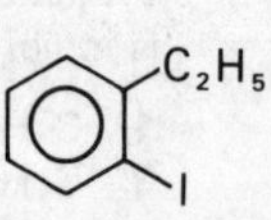

13 With which of the following would you expect PCl$_5$ to react *least* vigorously?

A C$_2$H$_5$OH   B C$_2$H$_5$CO$_2$H   C H$_2$O

D 2-methylphenol (benzene ring with CH$_3$ and OH substituents)

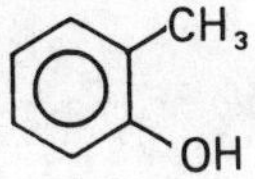

E benzene ring with CO$_2$H substituent

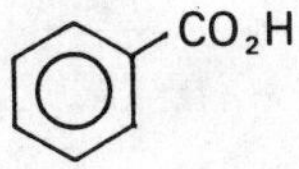

14 Which of the following in aqueous solution would liberate carbon dioxide from sodium carbonate solution?

A CH$_3$CO$_2$H   B C$_2$H$_5$OH   C benzene ring with OH (phenol)

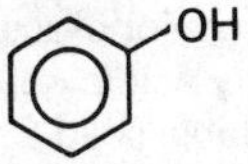

D benzene ring with OH and CH$_3$ E HOCH$_2$CH$_2$OH

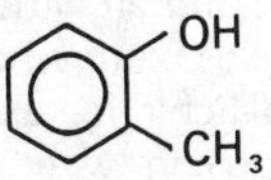

15 Which of the following is the formula for ethanoyl chloride?

A CH$_3$CH$_2$Cl   B CH$_3$OCl   C CH$_3$C(=O)Cl

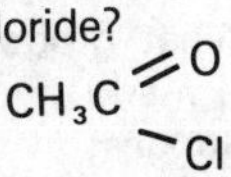

D CH$_3$C(OCl)(H)   E CH$_3$C(OH)(Cl)

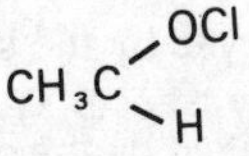

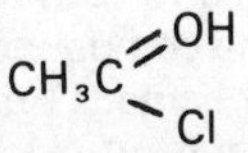

16 The organic ion resulting when propanoic acid reacts with sodium is

A C$_2$H$_5$CO$_2^-$ B CH$_3$CO$_2^-$ C C$_3$H$_7$CO$_2^-$
D C$_2$H$_5$CO$^-$ E C$_2$H$_6$CO$^-$

17 Which of the following reagents will replace —OH by a halogen atom?

A Cl$_2$ B SOCl$_2$ C HOCl   D I$_2$ E Br$_2$

18 In the preparation of ethyl ethanoate what is the function of the sodium carbonate in the purification?

A It is a drying agent
B It removes excess ethanol
C It separates the ester from the water
D It reacts with the calcium chloride used in the drying
E It reacts with any acids present

19 When nitric acid reacts with phenol which of the following is amongst the products?

A 2-nitrophenol (OH, NO$_2$)
B nitrobenzene (NO$_2$)
C phenyl nitrate (ONO$_2$)
D 1,2-dinitrobenzene (NO$_2$, NO$_2$)
E 3-nitrophenol (OH, NO$_2$)

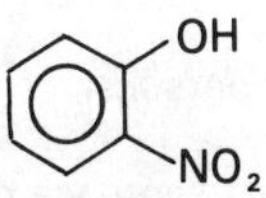

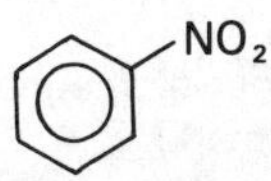

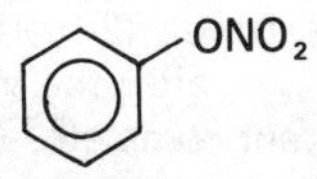

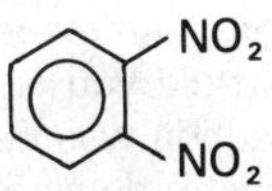

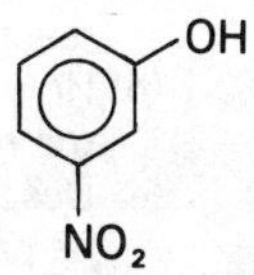

20 What colour results from adding a solution of iron(III) chloride to a solution of sodium ethanoate in water?

A red   B blue   C brown   D green   E purple

11B Structured questions

1 A white crystalline solid A, which burns with a smoky, luminous flame, has the formula $C_{13}H_{10}O_2$. A is refluxed with sodium hydroxide solution until it all dissolves, after which the reaction mixture is acidified with dilute hydrochloric acid. From this mixture two solids can be separated: a very faintly acidic solid B, which has the formula C_6H_6O, and an almost water-insoluble solid C with formula $C_7H_6O_2$. B gives a purple colour when treated with iron(III) chloride solution and reacts with benzoyl chloride and sodium hydroxide to give the compound A. C, though insoluble in water, dissolves freely in sodium hydroxide solution.

(a) What group of atoms is the iron(III) chloride test used to identify? (1 mark)
(b) Suggest a structural formula for B. (1 mark)
(c) What information about the compound A is conveyed by the statement that it burns with a smoky flame? (1 mark)
(d) What name is given to the type of reaction exemplified by the change of A into B and C? (1 mark)
(e) When B reacts with benzoyl chloride, what is the by-product and why is it not given off from the reaction mixture? (2 marks)
(f) Write a structural formula for C. (2 marks)
(g) Write a structural formula for A. (2 marks)

2 Propanone may be reduced by a water-free mercury/magnesium mixture (magnesium amalgam) to a substance called 'pinacol' which has the formula $C_4H_{10}O_2$. The propanone is found to dissolve mercury(II) chloride ($HgCl_2$) quite well. As soon as magnesium amalgam comes into contact with the propanone, a reaction ensues which begins slowly and may need warming to start it but which is vigorous when it is once started. The pinacol is liberated from the reaction mixture by the addition of water, heating gently for an hour, and subsequently adding benzene in which pinacol is soluble.

boiling points: propanone 56.2°C
                benzene   80.1°C
                pinacol   174.4°C

(a) Suggest a method of making magnesium amalgam, given a supply of dry magnesium and any reagents mentioned above. (2 marks)
(b) Assuming that you have some magnesium amalgam in a flask, with what extra apparatus would you fit the flask for the reduction? (1 mark)
(c) Describe in detail how you would perform the reduction (the 'vigorous reaction' mentioned above). (2 marks)
(d) How would you ensure, after the vigour of the reaction subsides, that the reaction was in fact complete? (1 mark)
(e) Describe in detail how the process described in the last sentence could be carried out to give a pure sample of pinacol. (4 marks)

3   It has been reported[9] that a method has been discovered of converting refuse from towns into oil. Refuse is predominantly cellulose and the method involves heating it with carbon monoxide and water at 500 K and 76 000 mmHg. The resulting oil is extracted with propanone or benzene and then recovered by distillation. The oil consists of alkanes and cycloalkanes with carbonyl and carboxylic acid side groups and has a sulphur content as low as 0.1%. The process operates at 40% efficiency.

cellulose $[C_6H_{10}O_5]_n$

(a) In what way does cellulose differ from starch? (2 marks)
(b) What physical method of analysis might have been used to decide that the oil consisted of alkanes and cycloalkanes with carbonyl and carboxylic acid side groups? (1 mark)
(c) What is the difference between a carbonyl group and a carboxylic acid group? (1 mark)
(d) The low sulphur content of this oil makes it especially useful as a fuel. What is the advantage of the low sulphur content? (2 marks)
(e) What disadvantage does the industrial scale use of benzene have which would make propanone a somewhat preferable choice? (2 marks)
(f) Why is it possible to use propanone for extracting the oil when propanone is miscible with water? (2 marks)

4   A piece of research work has been described[10] in which a number of varieties of broad bean were analysed for the presence of three coloured compounds in the seed coats. The three compounds all belonged to a class of compound called the 'anthocyanins' and are as follows:

cyanidin (yellow)

delphinidin (blue)

pelargonidin (red)

The seed coats of the broad beans were removed and extracted with a mixture of butan-1-ol and concentrated HCl. The extracts were subjected to paper chromatography and R_f values for the spots were compared with the R_f values for the three anthocyanins. The results were as follows:

	Cyanidin	Delphinidin	Pelargonidin	Variety: Broad Windsor	Three-fold White	Exhibition
R_f value obtained	0.44	0.30	0.58	0.49 0.29	no spots	0.47 0.27

The solvent used for chromatography was a mixture of ethanoic acid (30 volumes), water (10 volumes) and concentrated HCl (3 volumes).

(a) What is meant by the term 'R_f value'? (2 marks)
(b) Briefly describe how you would set about 'extracting' the seed coats (the outer skin of the broad bean seed). (2 marks)
(c) Chromatography paper has a layer of water 'adsorbed' on to its surface by hydrogen bonding with —OH groups in the molecules of the cellulose in the paper. Suggest a reason why the three anthocyanins should have different R_f values. (2 marks)
(d) Outline the practical details of the procedure by which the paper chromatography mentioned above could have been carried out, mentioning how the positions of the spots could have been detected. (4 marks)
(e) Would you expect the three anthocyanins to be solids, liquids or gases at room temperature? Justify your answer. (2 marks)
(f) Which anthocyanins were present in the seed coats of the three varieties of broad bean mentioned? Which spot gives the most difficulty in identifying it from its R_f value? How could the difficulty of identification be resolved? (3 marks)

SECTION 12

Equilibria

12A Fixed response items—equilibrium constant

In this section R is used in place of Lk, where L is the Avogadro constant and k the Boltzmann constant.

1   For the equilibrium $N_2O_4(g) \rightleftharpoons 2NO_2(g)$ which of the following statements is correct? (K_c is the equilibrium constant for molar concentrations, K_p is the equilibrium constant for partial pressures)

   A $K_c = K_p$ **B** $K_c = \sqrt{K_p}$ **C** $K_c = K_p^2$ **D** $K_c = \frac{1}{RT} \cdot K_p$
   E $K_c = RT.K_p$

2   The solubility product for magnesium hydroxide is 2×10^{-11} mol³ dm⁻⁹. This means that

   A $[Mg^{2+}] (\frac{1}{2}[OH^-])^2 = 2 \times 10^{-11}$ **B** $[Mg^{2+}] (2[OH^-])^2 = 2 \times 10^{-11}$
   C $[Mg^{2+}] [OH^-]^2 = 2 \times 10^{-11}$ **D** $\frac{1}{2}[Mg^{2+}] [OH^-]^2 = 2 \times 10^{-11}$
   E $[Mg^{2+}] [OH^-] = 2 \times 10^{-11}$

3 The units of K_c, the equilibrium constant in concentrations, are
   A mol dm^{-3} B mol^2 dm^{-3} C mol^{-1} dm^3 D mol^2 dm^{-6}
   E it depends on the particular reaction being considered

4 Hydrogen and nitrogen enter into equilibrium with ammonia according to the equation:

   $N_2(g) + 3H_2(g) \rightleftharpoons 2NH_3(g)$

   Suppose 0.1 mole of each of the three gases are mixed and allowed to reach equilibrium in a vessel of 1 dm^3 capacity. x moles of ammonia are converted into nitrogen and hydrogen. What is the concentration of hydrogen at equilibrium?

   A $0.1 + 0.66 x$ mol dm^{-3} B $0.1 + 1.5 x$ mol dm^{-3}
   C $0.1 + 0.5 x$ mol dm^{-3} D $0.1 + 0.1 x$ mol dm^{-3}
   E $0.1 + 0.75 x$ mol dm^{-3}

5 If 50 cm^3 of carbon dioxide and 150 cm^3 of hydrogen each at 760 mmHg are mixed together without reacting and the final pressure of the mixture is 760 mmHg, what is the partial pressure of the carbon dioxide?

   A 760 mmHg   B 380 mmHg   C 253 mmHg   D 190 mmHg
   E 152 mmHg

6 When hydrogen and iodine vapour react at 698 K an equilibrium is set up. There are several ways in which the equilibrium expression can be written and several correct equilibrium constants. Which of the following is *not* a correct statement about the equilibrium?

   A $\dfrac{4[HI]^2}{[H_2][I_2]} = $ constant   B $\dfrac{[H_2][I_2]}{[HI]^2} = K_c$

   C $\dfrac{[HI]}{[H_2][I_2]} = K_c$ D $K_p = \dfrac{p_{HI}^2}{p_{H_2} \times p_{I_2}}$

   E $K_p = \dfrac{p_{H_2} \times p_{I_2}}{p_{HI}^2}$

7-9 These items are concerned with the equilibrium whose equation is

$CH_3CH(OH)CH_3(g) \rightleftharpoons CH_3COCH_3(g) + H_2(g)$

Under the conditions of the experiment the reactants and products are all gaseous and the reaction is endothermic.

7 If the pressure on the system is increased the yield of propanone

   A increases
   B decreases but the yield of hydrogen remains the same
   C remains the same
   D decreases
   E increases but the yield of hydrogen remains the same

8 If the temperature is increased from 600 K to 700 K

    A the yield of propanone remains the same
    B the yield of hydrogen remains the same
    C the yield of propanone decreases
    D the yield of hydrogen decreases
    E the yield of propanone increases

9 If a catalyst of copper is employed, the yield of propanone

    A decreases because the reverse reaction is accelerated more than the forward one
    B increases because the hot copper absorbs the hydrogen
    C remains the same but the value of ΔH changes
    D increases because the forward reaction is accelerated more than the reverse one
    E remains the same, but the forward and the reverse reactions are both accelerated

10 Given that the equilibrium constant for the reaction

$$PCl_5(g) \rightleftharpoons PCl_3(g) + Cl_2(g)$$

is 0.0224 at 500 K and 33.3 at 760 K, which of the following statements is/are true?

    1 The reaction is endothermic.
    2 If the pressure on the system is kept constant and the temperature raised from 500 K to 760 K, the volume increases to an extent which cannot be explained merely by gas expansion
    3 The equilibrium constant in molar concentrations is numerically equal to the equilibrium constant in partial pressures

12B Fixed response items—acid-base equilibria

1-8 In these items, classify the reactions represented in the questions into one of the following categories:

    A oxidation only   **B** reduction only
    C oxidation and reduction   **D** acid-base
    E acid-base and redox

1 $Sn(s) + 2H_3O^+(aq) \rightarrow H_2(g) + Sn^{2+}(aq) + 2H_2O(l)$

2 $2C_2H_5OH(l) + 2Na(s) \rightarrow 2C_2H_5O^-Na^+(s) + H_2(g)$

3 $I_2(aq) + 2S_2O_3^{2-}(aq) \rightarrow 2I^-(aq) + S_4O_8^{2-}(aq)$

4 $Sn(OH)_4(s) + 2NaOH(aq) \rightarrow Na_2SnO_3(aq) + 3H_2O(l)$

5 $2SO_2(g) + O_2(g) \rightarrow 2SO_3(g)$

6 $IO_3^-(aq) + 5I^-(aq) + 6H_3O^+(aq) \rightarrow 3I_2(aq) + 9H_2O(l)$

7 $Mg(s) \rightarrow Mg^{2+}(aq) + 2e^-$

8 $H_2O_2 + 2I^- + 2H_3O^+ \rightarrow 4H_2O + I_2$

9-10 These items refer to the reaction whose equation is:

$NaH_2PO_4(aq) + NaOH(aq) \rightleftharpoons Na_2HPO_4(aq) + H_2O(l)$

in which the ions $H_2PO_4^-$ and HPO_4^{2-} form parts of the salts.

9 In this reaction, which ions do *not* undergo any chemical change (i.e. are spectator ions)?

A OH^-
B $H_2PO_4^-$ and OH^-
C HPO_4^{2-} and Na^+
D Na^+ and OH^-
E Na^+

10 What is the conjugate acid of OH^- in this reaction?

A Na^+ B H_2O C $H_2PO_4^-$ D HPO_4^{2-} E NaH_2PO_4

12C Fixed response items—pH and indicators

1-3 These items concern an experiment in which an indicator HIn of $pK_a = 4.5$ is added to a solution X and the colour thus obtained exactly matches a pair of test-tubes, one of which contains 4 drops of the acid colour of the indicator and the other contains 6 drops of the alkali colour of the indicator.

1 Which of the following is true of the solution X?

A Its pH is more than 4.5
B It is slightly alkaline
C Its pH is 4.5
D It is a buffer solution
E It is neutral

2 Which of the following is the predominant species in the 'alkali colour' of the indicator?

A H_2In B In^+ C HIn D $HOIn$ E In^-

3 If the formula relating pH to pK_a for the indicator is of the form

$pH = pK_a + \lg N$

which of the following is N?

A $[In^-]$
B $[HIn]$
C $\dfrac{[HIn]}{[In^-]}$
D $[In^-][HIn]$
E $\dfrac{[In^-]}{[HIn]}$

4-5 Classify the graphs mentioned in these items as being of one of the lettered types. In each case the quantity mentioned first is plotted on the vertical axis.

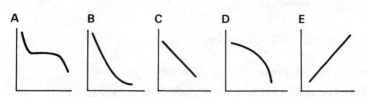

4 The volume of acid added to a certain volume of alkali plotted against the pH.

5 A graph of pH against lg[OH⁻] in the range [OH⁻] from 0.000 01 to 1 mol dm⁻³.

6 The pH of a 0.005 M solution of calcium hydroxide is

    A 14   **B** 13   **C** 12   **D** 11   **E** 10

7 On the following graph of pH change during an acid-alkali titration, in which of the lettered parts of the curve is the solution acting as a buffer solution?

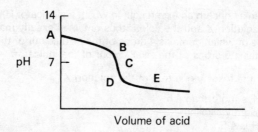

8 A graph like the one in the previous question could be obtained by reacting which of the following pairs of solutions together?

    A 0.1 M NaOH with 0.1 M HCl
    B 0.1 M Ca(OH)$_2$ with 0.1 M HCl
    C 0.1 M NH$_3$ with 0.1 M HCl
    D 0.1 M NH$_3$ with 0.1 M CH$_3$CO$_2$H
    E 0.1 M NH$_3$ with 0.1 M H$_2$SO$_4$

9 What is the pH of a buffer solution made by mixing 100 cm³ of 0.1 M methanoic acid (pK_a = 3.8) with 10 cm³ of 0.1 M sodium methanoate?

    A 2.8   **B** 3.7   **C** 3.8   **D** 3.9   **E** 4.8

10 The pH of a 0.01 M solution of a weak acid is 4.0. Which of the following is K_a for the acid in mol dm⁻³?

A 1×10^{-2} B 1×10^{-4} C 1×10^{-6}
D 1×10^{-8} E 1×10^{-10}

12D Structured questions

1. At a certain temperature the equilibrium constant K_c for the reaction:

    $POCl_3(g) \rightleftharpoons POCl(g) + Cl_2(g)$

    was 1.0 mol dm^{-3}. At equilibrium 40% of the original concentration of $POCl_3$ had decomposed.

    (a) What was the original concentration of $POCl_3$? (3 marks)
    (b) What would be the effect of increasing the pressure on the system? (2 marks)
    (c) The reaction as written is endothermic; what would be the effect of raising the temperature of the system? (2 marks)
    (d) What is the relationship between K_p and K_c for this reaction? (3 marks)

2. In an experiment involving the reaction:

    $C(s) + CO_2(g) \rightleftharpoons 2CO(g)$

    carbon dioxide is allowed to reach equilibrium with solid carbon at 1300 K and 40 atmospheres pressure. The equilibrium partial pressure of carbon dioxide is found to be 32 atmospheres.

    (a) Write an expression for the equilibrium constant in partial pressures, K_p. (2 marks)
    (b) Calculate a value for K_p using the data and give its units. (2 marks)
    (c) Calculate the mole fraction of carbon dioxide in the equilibrium mixture. (3 marks)
    (d) Calculate the average molecular mass of the gases in the mixture. (3 marks)

3. The equilibrium constant K_p for the reaction

    $H_2(g) + CO_2(g) \rightleftharpoons H_2O(g) + CO(g)$

    at 1260 K is 1.60. The reaction from left to right is exothermic. 1 dm³ each of hydrogen and carbon dioxide at 1260 K and 760 mmHg are mixed in a 2 dm³ vessel and allowed to reach equilibrium.

    (a) What is the partial pressure of carbon monoxide at equilibrium? (2 marks)
    (b) What would be the effect on the equilibrium partial pressure of carbon monoxide if the original pressures of hydrogen and carbon dioxide were double the values given above? (2 marks)
    (c) What would be the partial pressure of carbon monoxide at equilibrium if 2 dm³ of hydrogen and 2 dm³ of carbon dioxide at 1260 K and 760 mmHg were mixed, compressed into a vessel of 4 dm³ capacity and allowed to reach equilibrium? (2 marks)
    (d) What would be the effect of each of the following on
     (i) the rate of attainment of equilibrium and
     (ii) the partial pressure of carbon monoxide?
     1. putting in a catalyst
     2. lowering the temperature (4 marks)

4   The following table shows the pH of various aqueous solutions, all solutes being at 0.1 M concentration:

solute(s)	pH
hydrofluoric acid HF	2.1
hydrofluoric acid and sodium fluoride	3.3
phosphoric acid	1.5

(a) (i) What is the pH of 0.1 M HCl?
    (ii) HI is said to be a stronger acid than HCl. What is the pH of 0.1 M HI?
    (iii) What can be deduced about the strength of HF as an acid from the pH given? (3 marks)
(b) (i) What is the value of $[H^+]$ in 0.1 M phosphoric acid?
    (ii) If phosphoric acid ionizes
    $$H_3PO_4 \rightleftharpoons H^+ + H_2PO_4^-$$
    what is the value of $[H_2PO_4^-]$ in 0.1 M phosphoric acid?
    (iii) Calculate a value for K_a for phosphoric acid. (6 marks)
(c) What concentration of HCl would have a pH of 3.3? (2 marks)
(d) If 10 cm³ each of the solution of HCl with a pH of 3.3 and the mixture of pH 3.3 mentioned in the table are treated with a few drops of M NaOH, what would you expect to happen to the pH of the solutions? (No calculation is expected here.) Explain why the two solutions behave differently. (6 marks)
(e) The indicator methyl yellow has pK_a 3.5. It is red at low pH and yellow at higher pH. What colour would it be in each of the solutions mentioned in the table? (3 marks)

SECTION 13

Organic Chemistry 3

13A Fixed response items—amines and carbonyl compounds

1 Which of the following ions results from the action of hydrochloric acid on butylamine?

A $C_3H_7NH_2^+$ B $C_4H_9NH_2^+$ C $C_4H_9NH_4^+$
D $C_3H_7NH_3^+$ E $C_4H_9NH_3^+$

2 Which of the following is the *least* soluble in water?

A ammonia   B propanone   C phenylamine
D butylamine   E ethanal

3 Which of the following would *not* give a deep blue solution when treated with copper(II) sulphate solution?

A   B NH_3 C $C_4H_9NH_2$
D CH_3NH_2 E $NH_2CH_2CO_2H$

4 Which of the following is the formula for the main organic product when butylamine reacts with nitrous acid?

A C_4H_9OH B $C_3H_7CO_2H$ C $C_4H_9NO_2$
D $C_4H_9CO_2H$ E $C_4H_8(OH)_2$

5 Which of the following are optimum temperature conditions for making benzenediazonium chloride from phenylamine?

A −10°C to 0°C   B 5°C to 10°C   C 15°C to 25°C
D 30°C to 40°C   E 45°C to 50°C

6 Which of the following is/are used in the making of benzenediazonium chloride from phenylamine?

1 sodium nitrite
2 hydrochloric acids
3 phenol

7-10 In each of these items, classify the substance mentioned into one of the lettered categories.

A ketone   B amine   C aldehyde   D ester
E carboxylic acid

7 CH_3—⌬—CO_2H

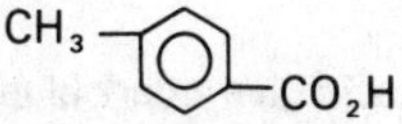

8 —CH_2NH_2

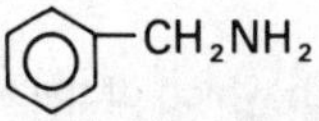

9

10 ⌬=O (cyclohexanone)

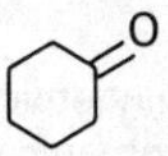

11 When carbonyl compounds react with Fehling's solution, what is the red precipitate which is formed?

A copper   B an organic acid   C copper(I) oxide
D an organic dye   E copper(II) oxide

12 Which of the following will react with Fehling's solution giving a red precipitate?

A C_3H_7CHO B $C_2H_5COCH_3$ C C_3H_7OH
D $CH_3OC_2H_5$ E C_6H_5OH

13 Which of the following would be required when oxidizing propan-2-ol to give propanone?

    1 sodium hydroxide   **2** sulphuric acid   **3** sodium dichromate

14 Which of the following techniques would be involved in obtaining a sample of ethanal from ethanol?

    1 refluxing   **2** recrystallization   **3** distillation

15 Which of the following represents the formula of the product of the reaction between propanone and 2,4-dinitrophenylhydrazine?

A $CH_3-C(=O)-NH-NH-C_6H_3(NO_2)_2$

B $CH_3-CH(CH_3)-N=N-C_6H_3(NO_2)_2$

C $CH_3-C(CH_3)=N-NH-C_6H_3(NO_2)(NO)$

D $CH_3-C(CH_3)-O-NH-C_6H_3(NO_2)_2$

E $CH_3-C(OH)-CH_2-NH-NH-C_6H_3(NO_2)_2$

16 When aminoethanoic acid (glycine) reacts with nitrous acid the main product is

    A $HOCH_2CO_2H$ **B** $HOCH_2CO_2NH_2$ **C** $NH_2CH_2CO_2NH_4$
    D $NH_2CH_2CONH_2$ **E** $HO_2CCH_2CO_2H$

17 With which of the following will aminoethanoic acid give a dark blue colour?

    A HNO_2 **B** Cu   **C** Cu^{2+} **D** HCl   **E** NaOH

18 When aminoethanoic acid reacts with sodium hydroxide which of the following organic ions results?

    A $NH_2CH_2CO_2^-$ **B** $NH_2CH_2CO^-$ **C** $\overset{+}{N}H_3CH_2CO_2^-$
    D $\overset{+}{N}H_3CH_2CO_2H$ **E** $\overset{+}{N}H_4CH_2CO_2H$

19 Which of the following is/are amongst the products when ethanamide reacts with nitrous acid?

    1 CH_3CO_2H **2** N_2 **3** C_2H_5OH

20 When aminoethanoic acid reacts with hydrochloric acid, which of the following organic ions results?

    A $NH_2CH_2CO_2^-$ **B** $\overset{+}{N}H_3CH_2CO_2^-$ **C** $\overset{+}{N}H_3CH_2CO_2H$
    D $\overset{+}{N}H_3CH_2CONH_2$ **E** $HOCH_2CO_2^-$

3B Fixed response items—proteins

1. Which of the following would *not* show a colour with ninhydrin?

    A glycine   B alanine   C glucose   D saliva   E perspiration

2. Which of the following best explains why a protein successfully gives an X-ray diffraction pattern?
    A Methyl groups occur at regular intervals along a protein chain
    B The peptide groups occur at regular intervals along the helical chain of the protein
    C The sulphur atoms in some amino acid units link one part of the protein chain to another
    D The —C=O groups of the peptide link are always orientated in the same direction
    E The 'rods' in a 'rod-like' protein are regularly arranged like the ions in an ionic crystal

3. Along the main helical chain of a protein, the sequence of atoms is:

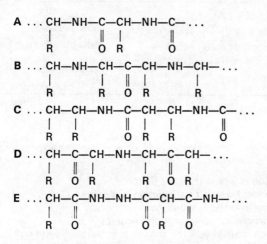

4. Which of the following might help to identify the amino acids on a paper chromatogram?
    1 Compare their R_f values with those of various known amino acids
    2 Spray the chromatogram with ninhydrin
    3 Develop the paper chromatogram in two directions at right angles and compare R_f values with those of known amino acids

5. When Sanger was doing his classic work on the structure of insulin, which of the following techniques were used?
    1 gas chromatography
    2 fractional distillation
    3 two-directional paper chromatography

6   In paper chromatography of amino acid and polypeptide mixtures, the function of 1-fluoro-2,4-dinitrobenzene is

   1   to identify the amino acid at one end of a polypeptide chain
   2   to distinguish between free amino acids and dipeptides
   3   to distinguish between a protein and a polypeptide

7   Which of the following can affect the activity of a particular enzyme?

   1   adding 0.1 M sodium hydroxide solution to the substrate
   2   raising the temperature of the substrate
   3   adding 0.1 M hydrochloric acid to the substrate

8-10   These items concern the following experiment:

A few drops of the acid-alkali indicator bromothymol blue are added to each of six beakers containing various aqueous liquids as follows:

	1	2	3	4	5	6
Colour	yellow	blue	yellow	yellow	yellow	blue
Contents	buffer pH 7	buffer pH 8	distilled water	distilled water + urea (NH_2CONH_2)	distilled water + enzyme	distilled water + urea + enzyme

8   Which of the following conclusions is *not* correct from the experimental results?

   A   Bromothymol blue is yellow in acid solution
   B   Urea only reacts with water if the solution is alkaline
   C   A solution of urea in water is either neutral or acidic
   D   The indicator changes colour over a pH range of 7 to 8
   E   The solution would remain yellow if the water in beaker 6 were replaced by a buffer of pH 7

9   Which of the following reactions is most likely to be going on in beaker 6?

   A   Oxidation of urea giving nitric acid as one of the products
   B   Reduction of urea giving nitrogen dioxide as one of the products
   C   Addition of water molecules to the carbonyl group in urea
   D   Hydrolysis of urea giving ammonia as one of the products
   E   Hydrolysis of urea giving nitrous acid as one of the products

10   The contents of beaker 6 were boiled and then allowed to cool. The solution was restored to pH 7 by cautiously adding very dilute hydrochloric acid. When a further 5 minutes had elapsed, the solution was still yellow. Which of the following explanations is most likely?

A The boiling had converted the enzyme into its constituent amino acids
B Dissolved air had been expelled from the solution by boiling
C The shape of the enzyme molecule had been altered by boiling
D The enzyme works much faster at higher temperatures so no urea was left
E The reaction is very slow at pH 7, more time is needed before the colour will change

3C Structured questions

1 This question concerns the following instructions:[11]

Small-scale preparation of bromobenzene

The experiment must be carried out in a fume cupboard, and take great care not to breathe in the vapour of either benzene or bromine. Benzene is highly toxic and must only be used under supervision.

Set up the apparatus shown in the diagram with 6 cm³ of benzene and 0.2 g of iron filings in the flask. Run 3 cm³ of bromine slowly into the flask, and *slowly*

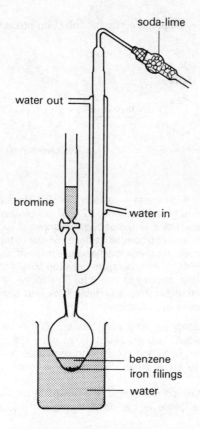

raise the temperature of the water-bath to 70°C. Maintain this temperature until no more hydrogen bromide is evolved.

Remove the flask, cool it in cold water and pour the mixture into a separating funnel.

(a) Using the data provided, work out whether bromine or benzene is in excess, showing the steps in your working clearly.

	benzene	bromine
Relative molecular mass	78	160
Density (g cm^{-3})	0.9	3.12

(3 marks)

(b) Write an equation for the reaction between bromine and benzene (2 marks)

(c) Mention *two* reasons why a similar method would not be suitable for making 1-bromohexane. (2 marks)

(d) There would be three principal steps in the purification of the bromobenzene prepared by this method. Outline them, mentioning the names of any reagents required and making it clear what each stage is supposed to do. (3 marks)

2 Cyclohexanone may be prepared by the following procedure:

Scheme

Cyclohexanol $\xrightarrow{Na_2Cr_2O_7/H_2SO_4}$ Cyclohexanone

Method

Dissolve 25 g of sodium dichromate crystals in 125 cm^3 of water in a large beaker (500 cm^3) and add carefully, with stirring, 22 g of concentrated sulphuric acid. Allow the mixture to cool. Place 12.5 g of cyclohexanol in a 250 cm^3 conical flask and add the dichromate solution to it all at once. Shake to mix and monitor the temperature of the mixture. As considerable heat is evolved, the temperature will rise quite rapidly. Maintain the temperature between 55°C and 60°C. When the reaction is over, allow the mixture to stand for 30 minutes. Add 100 cm^3 more water to the mixture, transfer it to a suitable flask and distil. Two layers are collected; the upper one is cyclohexanone.

(a) Why is it necessary to use a *large* beaker? (first sentence) (1 mark)

(b) How exactly would you go about maintaining the temperature of the mixture between 55°C and 60°C (sixth sentence) (2 marks)

(c) How would you know when the reaction is over? (seventh sentence) (1 mark)

(d) During the distillation mentioned, if a thermometer were incorporated in its conventional place in the distillation apparatus, what temperature would it register? (The boiling point of cyclohexanone is 155°C.) (2 marks)

(e) Continue the 'method' to indicate how to get a pure, dry sample of cyclohexanone. (4 marks)

SECTION 14

Rate of reaction

14A Fixed response items

In this section R is used in place of Lk, where L is the Avogadro constant and k the Boltzmann constant.

1. The units of rate of reaction
   - **A** vary according to the rate equation
   - **B** are $mol^{-1}\ dm^6\ s^{-1}$
   - **C** are $mol\ dm^{-3}$
   - **D** are $mol\ dm^{-3}\ s^{-1}$
   - **E** are $mol\ s^{-1}$

2. The units of rate constant
   - **A** vary according to the rate equation
   - **B** are $mol\ dm^{-6}\ s$
   - **C** are $mol^{-1}\ dm^3\ s^{-1}$
   - **D** are $mol\ s^{-1}$
   - **E** are $mol\ dm^{-3}\ s^{-1}$

3. For a reaction involving a substance X, doubling the concentration of X changes the rate of the reaction from 7.0×10^{-6} to 2.8×10^{-5} $mol\ dm^{-3}\ s^{-1}$. What is the order of the reaction with respect to X?

   A 1   **B** 2   **C** 3   **D** 4   **E** 5

4. For the reaction $2NO + Cl_2 \rightarrow 2NOCl$ the rate equation is:

   rate = $k\ [NO]^2[Cl_2]$

   If the concentrations of NO and Cl_2 at the start of the reaction are both $0.01\ mol\ dm^{-3}$ what is the rate of the reaction when the concentration of NO is $0.005\ mol\ dm^{-3}$?

   - **A** $k \times 0.005 \times 0.0025$
   - **B** $k \times 0.005 \times 0.005$
   - **C** $k \times 0.005^2 \times 0.0025$
   - **D** $k \times 0.005 \times 0.0075$
   - **E** $k \times 0.005^2 \times 0.0075$

5. The activation energy of a gas reaction depends on the nature of the bonds broken in the rate-determining step. For a reaction involving the addition of a halogen to an alkene, which of the following is the most reasonable estimate of the activation energy in $kJ\ mol^{-1}$?

   A 2000   **B** 200   **C** 10   **D** -200 **E** -2000

6-8 These items refer to the following information:
The reaction

$$NO(g) + O_3(g) \rightarrow NO_2(g) + O_2(g)$$

is a second order reaction overall. It takes place in a single step.

6 What would be the effect on the rate of the reaction of halving the pressure of the NO whilst keeping the pressure of O_3 constant?

- **A** it would not change
- **B** it would decrease by half
- **C** it would decrease by three quarters
- **D** it would decrease by a quarter
- **E** it would decrease by one eighth

7 If the concentrations of the NO and of the O_3 were both doubled, what would be the effect on the initial rate of the reaction?

- **A** it would not change
- **B** it would increase by a quarter
- **C** it would double
- **D** it would increase to three times its original value
- **E** it would increase to four times its original value

8 Which of the following is a possible rate equation for the reaction?

- **A** rate = $k[NO]^2$
- **B** rate = $k[O_3]^2$
- **C** rate = $k[NO][O_3]$
- **D** rate = $k[NO][O_3]^0$
- **E** rate = $k[NO]^0[O_3]$

9-10 These items concern the collision theory of reaction kinetics.

9 If in a second order gas reaction at 500°C the number of collisions per second between molecules is 1×10^{33}, which of the following is the best estimate of the number of *effective* collisions per second?

- **A** 1×10^{30}
- **B** 1×10^{28}
- **C** 1×10^{20}
- **D** 1×10^5
- **E** 1×10^2

10 The rate constants of many second order gas reactions at room temperature are similar. Which of the following comments is *untrue*?

- **A** the rates of these reactions are similar at similar reactant concentrations
- **B** the rates of these reactions will change with temperature
- **C** the activation energies of these reactions are similar
- **D** the rate constants are independent of pressure
- **E** most of the collisions between gas molecules result in reaction

11 The mechanism of the reaction

$$5HBr + HBrO_3 \rightarrow 3Br_2 + 3H_2O$$

is thought to involve the slow step

$$HBr + HBrO_3 \rightarrow HOBr + HBrO_2$$

Which of the following rate equations is consistent with this?

A rate = $k[HBr]^5[HBrO_3]$
B rate = $k[HBr]^5$
C rate = $k[HBr]^2$
D rate = $k[HBrO_3]^2$
E rate = $k[HBr][HBrO_3]$

12 A knowledge of k, the rate constant for a reaction, at a number of temperatures can be made to give information about

1. the heat of reaction
2. the equilibrium constant of the reaction
3. the activation energy of the reaction

13-16 These items concern the hydrolysis of benzene diazonium chloride which follows the equation:

$$C_6H_5N_2^+Cl^-(aq) + H_2O(l) \rightarrow C_6H_5OH(aq) + N_2(g) + HCl(aq)$$

Two students investigated the rate of this hydrolysis at 328 K using the following apparatus:

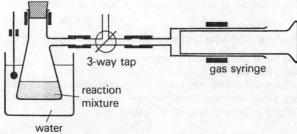

The benzene diazonium chloride was prepared in the reaction vessel using 0.7 g phenylamine, 2.2 cm³ concentrated HCl and a solution of 0.5 g of sodium nitrite in 75 cm³ of water. The flask was then placed in position in a beaker of warm water and allowed to warm up. The temperature was adjusted to 328 K and kept at that throughout the experiment.

The nitrogen liberated during the experiment was then collected in the gas syringe and the results were as follows:

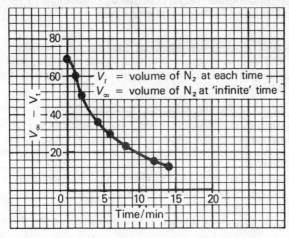

63

13 The purpose of the 3-way tap in this apparatus was

    A to ensure atmospheric pressure inside the apparatus
    B to allow nitrogen to escape while the apparatus was being brought up to 328 K
    C to control the rate of collection of nitrogen
    D to allow for thermal expansion of the nitrogen
    E to enable more than one gas syringe to be used

14 The volume of nitrogen liberated after 'infinite' time

    A depends on the concentration of benzene diazonium chloride when collection was started
    B depends on the rate of the reaction when collection was started
    C would have been different if 100 cm^3 of water had been used instead of 75 cm^3
    D would have been different if 0.6 g of sodium nitrite had been used instead of 0.5 g
    E was collected at one atmosphere pressure and 328 K

15 Which of the following best represents the order of the reaction?

    A first order with respect to water
    B first order overall
    C second order with respect to benzene diazonium chloride
    D first order with respect to benzene diazonium chloride
    E second order with respect to water

16 The small irregularity at the beginning of the graph could have been caused by

    A reading the volume too late
    B starting readings at too high a temperature
    C the syringe plunger 'sticking' just before the reading at 1 minute was taken
    D overestimating the volume collected at 1 minute
    E some air being in the syringe at the start

17-20 Consider the following lettered methods of 'following' a chemical reaction with a view to finding the order and rate constant, then choose the method which you consider most appropriate for each of the reactions mentioned in items 17–20.

    A measurements on the colour of the mixture
    B measurements on the volume of gas evolved from the mixture
    C measurements on the electrical conductance of the mixture
    D measurements on the time taken for a certain amount of precipitate to form
    E measurements involving 'sampling' the solution and titrating the samples with standard acid or alkali

17 The reaction between a solution of sodium carbonate and dilute acid

18 The reaction between hydrogen peroxide solution and potassium iodide solution

19 The reaction between potassium dichromate and ethanol in very acid solution

20 The reaction between barium chloride solution, sulphamic acid (NH_2SO_3H) and water to give solid barium sulphate and ammonium chloride solution

4B Structured questions

1 A number of experiments were performed on the rate of the reaction between hydrogen and nitrogen monoxide at 1100 K

$$2H_2(g) + 2NO(g) \rightarrow 2H_2O(g) + N_2(g)$$

The technique was to mix different molar proportions of the reacting gases and to make an estimate of the initial rate of the reaction. The results for various experiments were as follows:

Experiment number	Initial partial pressure NO (mmHg)	H_2	Initial rate mmHg s^{-1}
1	12	2	0.33
2	12	4	0.67
3	12	6	1.00
4	2	12	0.055
5	4	12	0.22
6	6	12	0.60

(a) What physical measurements would you make in order to find out how fast this reaction was going just after the reactants were mixed? (2 marks)
(b) What effect does doubling the concentration of nitrogen monoxide have on the rate of the reaction? (1 mark)
(c) What is the order of the reaction with respect to nitrogen monoxide? (1 mark)
(d) What effect does doubling the concentration of hydrogen have on the rate of the reaction? (1 mark)
(e) What is the order of the reaction with respect to hydrogen? (1 mark)
(f) Write the rate equation for this reaction. (1 mark)
(g) Calculate a value for k, the rate constant for this reaction. (1 mark)
(h) What can be deduced about the mechanism of the reaction from these data? (2 marks)

2 Potassium dissolves in liquid ammonia to give a dark blue solution. The potassium gradually reacts with the ammonia according to the equation

$$K + NH_3 \rightarrow KNH_2 + \tfrac{1}{2}H_2$$

The colour of the solution fades as the reaction proceeds. Measurements are taken which are proportional to potassium concentration at a number of times. Some results are as follows:

Time/hrs	Measurement
0.5	0.82
1.0	0.80
4.5	0.66
7.5	0.55
10.6	0.48
21.5	0.25

(a) What measurement would it be appropriate to make in order to 'follow' this reaction? (2 marks)
(b) What can be said of the concentration of the ammonia during this experiment? (1 mark)
(c) Plot suitable graph(s) and determine the apparent order of the reaction. (4 marks)
(d) (i) What additional information would be required in order to write a complete rate equation for the reaction?
(ii) Why would it be difficult to acquire this information? (3 marks)

3 The decomposition of nitrogen pentoxide N_2O_5 was studied at a number of different temperatures. The rate constant k at each temperature was obtained and the graph of $\ln k$ against $1/T$ is shown.

$$\ln k = C - \frac{E_a}{R} \cdot \frac{1}{T} \qquad R = 8.4 \text{ J K}^{-1} \text{ mol}^{-1}$$

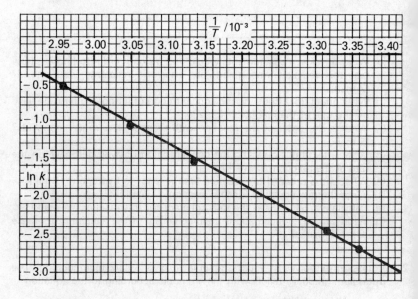

(a) Explain in words how the equation given means that reaction rate increases with temperature. (3 marks)

(b) Find the slope of the graph. (3 marks)
(c) Use this slope to obtain a value for the activation energy of the reaction. (4 marks)

SECTION 15

Voltaic cells

15A Fixed response items

In this section R is used in place of Lk, where L is the Avogadro constant and k the Boltzmann constant.

1-5 These items concern the following standard electrode potentials:

$Zn^{2+}(aq)$ | $Zn(s)$	-0.76 V
$Fe^{2+}(aq)$ | $Fe(s)$	-0.44 V
$Cd^{2+}(aq)$ | $Cd(s)$	-0.40 V
$Co^{2+}(aq)$ | $Co(s)$	-0.28 V
$Ni^{2+}(aq)$ | $Ni(s)$	-0.25 V
$Sn^{2+}(aq)$ | $Sn(s)$	-0.14 V

1 Which of the following is able to reduce $Co^{2+}(aq)$ to $Co(s)$?

   A $Ni^{2+}(aq)$ **B** $Ni(s)$ **C** $Cd^{2+}(aq)$ **D** $Sn(s)$ **E** $Cd(s)$

2 Which of the following can be reduced by $Ni(s)$?

   A $Sn^{2+}(aq)$ **B** $Cd^{2+}(aq)$ **C** $Fe^{2+}(aq)$ **D** $Zn^{2+}(aq)$ **E** $Co^{2+}(aq)$

3 What is the standard e.m.f. of the following cell?

   $Cd(s)$ | $Cd^{2+}(aq)$ ┊ $Ni^{2+}(aq)$ | $Ni(s)$

   A $+0.65$ V   **B** $+0.25$ V   **C** $+0.15$ V   **D** -0.25 V   **E** -0.65 V

4 The standard e.m.f. of the cell

   $Sn(s)$ | $Sn^{2+}(aq)$ ┊ $Zn^{2+}(aq)$ | $Zn(s)$

is -0.62 V. If the concentration of both the zinc ions and the tin ions were changed to 0.1 M what would happen to the e.m.f. of the cell?

   A It would remain unchanged
   B It would have a larger negative value than $E^{\ominus}_{cell}$
   C It would have a smaller negative value than $E^{\ominus}_{cell}$
   D It is impossible to say without further data
   E It would take on a positive value

5 The standard e.m.f. of the following cell is − 1.41 V:

Ni(s) | Ni²⁺(aq) ┊ Al³⁺(aq) | Al(s)

What is the standard e.m.f. of the aluminium electrode?

A + 1.66 V   B + 1.16 V   C + 1.41 V   D − 1.16 V   E − 1.66 V

6-10 These items concern the following electrodes:

I_2(aq), $2I^-$(aq) | Pt	+ 0.54 V
Fe^{3+}(aq), Fe^{2+}(aq) | Pt	+ 0.77 V
Br_2(aq), $2Br^-$(aq) | Pt	+ 1.09 V
[MnO_2(s) + $4H^+$], [Mn^{2+}(aq) + $2H_2O$(l)] | Pt	+ 1.23 V
Ce^{4+}(aq), Ce^{3+}(aq) | Pt	+ 1.70 V

6 Which of the following values seems most reasonable for the standard e.m.f. of this electrode?

Cl_2(aq), $2Cl^-$(aq) | Pt

A + 0.25 V   B + 0.42 V   C + 0.63 V   D + 1.01 V   E + 1.36 V

7 Which of the following is capable of reducing Fe^{3+}(aq) to Fe^{2+}(aq)?

A Br_2(aq)   B Mn^{2+}(aq)   C Ce^{3+}(aq)   D I_2(aq)   E I^-(aq)

8 What is the standard e.m.f. of this cell?

Pt | $2I^-$(aq), I_2(aq) ┊ Ce^{4+}(aq), Ce^{3+}(aq) | Pt

A + 2.14 V   B + 1.16 V   C + 1.70 V   D − 1.16 V   E − 2.14 V

9 The e.m.f. of the following cell is + 1.46 V.

Pt | Ti^{3+}(aq), Ti^{2+}(aq) ┊ Br_2(aq), $2Br^-$(aq) | Pt

What is the e.m.f. of the titanium electrode?

A + 1.46 V   B + 1.09 V   C + 0.37 V   D − 0.37 V   E − 1.46 V

10 Which of the following is capable of oxidizing potassium bromide solution?

A iron(III) chloride solution
B iron(II) sulphate solution
C cerium(III) sulphate solution
D iodine solution
E cerium(IV) sulphate solution

15B Structured questions

1 The standard e.m.f. of the electrode

Fe^{3+}(aq), Fe^{2+}(aq) | Pt

is the e.m.f. of a cell made up of this electrode and a standard hydrogen electrode joined together.
(a) Write the conventional cell diagram for such an arrangement. (2 marks)
(b) Draw an actual experimental arrangement for such a cell, labelling the diagram fully to show the standardizing conditions and the exact contents of each electrode. (8 marks)

2. There has been much debate at various times about the formula of the mercury(I) ion, whether it should be Hg^+ or whether it is Hg_2^{2+}, with a covalent bond between two mercury atoms. One piece of evidence is based on the number of electrons transferred when mercury atoms become mercury(I) ions, i.e. on the value of z in the formula

$$E = E^{\ominus} + \frac{RT}{zF} \ln C$$

where C is the concentration of mercury(I) ions.
A concentration cell is set up as follows:

Hg | mercury(I) nitrate ¦ mercury(I) nitrate | Hg
        0.05 M   ¦   0.005 M

for which the e.m.f. is 0.029 V at 298 K.
$R = 8.4$ J mol^{-1} K^{-1}; $F = 96\,500$ coulombs

(a) Using the formula given above, derive a formula for the e.m.f. of the concentration cell shown, in terms of R, T, z, F and concentrations. (4 marks)
(b) Hence calculate a value of z. (2 marks)
(c) What conclusion does this suggest about the formula of the mercury(I) ion? (1 mark)
(d) Sketch an apparatus which might be used to contain a concentration cell such as the one mentioned above. (3 marks)

3   The potentials of some standard electrodes are as follows:

Co^{2+}(aq) | Co(s)   −0.28 (cobalt)
Zn^{2+}(aq) | Zn(s)   −0.76 (zinc)
Pb^{2+}(aq) | Pb(s)   −0.13 (lead)
Sr^{2+}(aq) | Sr(s)   −2.89 (strontium)
Ba^{2+}(aq) | Ba(s)   −2.90 (barium)
Ni^{2+}(aq) | Ni(s)   −0.25 (nickel)

(a) Which of the metals mentioned in the list is the best reducing agent? (2 marks)
(b) Suppose it were required to convert some zinc metal into zinc ions, choose one metal or solution of metal ions from the list which would do this. (2 marks)
(c) The electrode Cr^{3+}(aq), Cr^{2+}(aq) | Pt has a standard redox potential of −0.41 V. Choose from the above list one metal or solution of metal ions which might be expected to convert Cr^{2+} ions into Cr^{3+} ions. (2 marks)
(d) Supposing that, having made a correct choice in (c) using electrode potentials, on trying the experiment nothing happens. What could be the cause of the failure? (2 marks)

(e) Predictions like the one in (c) often work despite one important condition attached to the use of $E^\ominus$ values which is seldom observed in simple qualitative work involving metals and their ions. What is this condition? (2 marks)

4 Use the Ellingham diagram below to answer the questions.

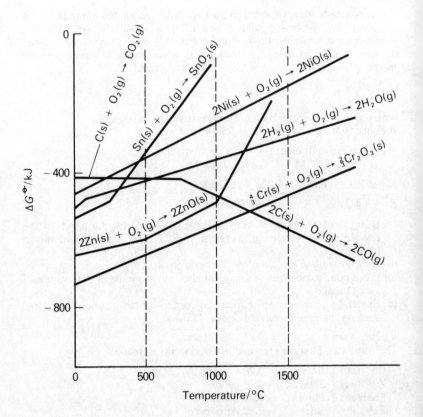

(a) Above what temperature should carbon be able to reduce chromium oxide? (1 mark)
(b) Which equations correspond to reactions for which $\Delta S^\ominus_{298}$ is positive? (2 marks)
(c) In which reactions do neither the metal nor its oxide melt over the temperature ranges shown? (2 marks)
(d) Which oxide shown on the diagram would decompose at the lowest temperature if heated alone? (1 mark)
(e) Which metal oxides could be reduced by hydrogen at 500°C? (2 marks)
(f) In what way, if at all, does $\Delta H_f^\ominus$ vary with temperature? (1 mark)
(g) Estimate a value for $\Delta H_f^\ominus$ of CO_2 using the diagram. (1 mark)

5 The reaction

$$CaCO_3(s) \rightleftharpoons CaO(s) + CO_2(g)$$

has $\Delta H^\ominus = +177$ kJ mol^{-1}. At 500 K, K_p for this reaction is 6.3×10^{-11} atm. $R = 8.2 \times 10^{-3}$ kJ mol^{-1} K^{-1}.

(a) Write an expression for K_p for this reaction. (1 mark)
(b) Calculate the temperature at which $K_p = 1$ atm. (3 marks)
(c) What, in terms of the equation, actually happens to the calcium carbonate if it is heated to, or beyond, this temperature? (1 mark)
(d) (i) What is the value of $\Delta G^\ominus$ at 500 K?
    (ii) What is the value of $\Delta G^\ominus$ when $K_p = 1$?
    (iii) Explain qualitatively in terms of standard entropy change $\Delta S^\ominus$ why $\Delta G^\ominus$ changes in this way when the temperature is raised. (5 marks)

6 Iron(II) ions Fe^{2+}(aq) can react with silver ions Ag^+(aq). Standard electrode potentials are:

Fe^{3+}(aq), Fe^{2+}(aq) | Pt; $E^\ominus = +0.77$ V
Ag^+(aq) | Ag(s); $E^\ominus = +0.80$ V

(a) Write a cell diagram for a cell incorporating these electrodes with the silver electrode written on the right. (1 mark)
(b) Write an ionic equation for the reaction which would occur if the metal parts of the two electrodes were connected together by a conductor. (1 mark)
(c) (i) Calculate the e.m.f. of the cell.
    (ii) From this e.m.f. calculate $\Delta G^\ominus$ for the reaction. ($F = 96\ 500$ C mol^{-1}) (3 marks)
(d) Thermodynamic data for the formation of the ions is as follows:

Ion	$\Delta G^\ominus_{f,298}$/kJ mol^{-1}	$\Delta H^\ominus_{f,298}$/kJ mol^{-1}
Fe^{2+}(aq)	-84.9	-87.8
Fe^{3+}(aq)	-9.7	-47.6
Ag^+(aq)	$+77.1$	$+105.6$

Use them to calculate another value for $\Delta G^\ominus_{298}$ for the reaction. (2 marks)
(e) Using the data, calculate a value for $\Delta H^\ominus_{298}$ for the reaction. (2 marks)
(f) Explain with the aid of the ionic equation in (b) why $\Delta H^\ominus_{298}$ and $\Delta G^\ominus_{298}$ have significantly different values for this reaction. (1 mark)

SECTION 16

The Periodic Table 4: the transition elements

16A Fixed response items

1 If the hydrated chromium(III) ion is $Cr(H_2O)_6^{3+}$, which of the following represents the stability constant for the complex $Cr(H_2O)_4en^{3+}$ where 'en' is 1,2-diaminoethane, $NH_2CH_2CH_2NH_2$?

A $K_1 = \dfrac{[Cr(H_2O)_4en^{3+}]}{[Cr(H_2O)_6^{3+}][en]}$
B $K_1 = \dfrac{[Cr(H_2O)_4en^{3+}][en]}{[Cr(H_2O)_6^{3+}]}$
C $K_1 = \dfrac{[Cr(H_2O)_6^{3+}]}{[Cr(H_2O)_4en^{3+}][en]^2}$
D $K_1 = \dfrac{[Cr(H_2O)_6^{3+}][en]^2}{[Cr(H_2O)_4en^{3+}]}$
E $K_1 = \dfrac{[Cr(H_2O)_4en^{3+}]}{[Cr(H_2O)_6^{3+}][en]^2}$

2-5 These items concern a complex of cobalt with the name potassium trioxalatocobaltate(III) whose formula is $K_3[Co(C_2O_4)_3]\,3H_2O$. The oxalate ion is $C_2O_4^{2-}$.

2 What is the oxidation number of cobalt in this complex?

   A +6   B −6   C +3   D −3   E +2

3 The oxalate ligand is

   A monodentate   B bidentate   C tridentate
   D quadridentate   E hexadentate

4 The overall charge on the complex ion is

   A −1   B +1   C −2   D −3   E +3

5 The lone pairs of electrons which form the bonds between the ligands and the metal ion in this complex are associated with

   A all the oxygen atoms in each oxalate ion
   B one oxygen atom in each oxalate ion
   C two of the oxygen atoms in each oxalate ion
   D three of the oxygen atoms in each oxalate ion
   E the carbon atoms in each oxalate ion

6-10 These items concern a complex ion of chromium with the formula $CrF_4(H_2O)_2^{-}$

6 The oxidation number of chromium in this complex is

   A −1   B +1   C +2   D +3   E +6

The coordination number of chromium in this complex is

A 2   B 3   C 4   D 5   E 6

In all probability the shape of this complex would be

A square planar   B tetrahedral   C rectangular
D double triangular   E octahedral

Which of the following is isomeric with

$$\begin{array}{c} H_2O \diagdown \quad \diagup F \\ \quad Cr - F \quad ? \\ H_2O \diagup \quad | \diagdown F \\ F \end{array}$$

A
$$\begin{array}{c} H_2O \\ F \diagdown | \diagup F \\ \quad Cr \\ H_2O \diagup | \diagdown F \\ H_2O \end{array}$$

B
$$\begin{array}{c} F \\ F \diagdown | \diagup H_2O \\ \quad Cr \\ H_2O \diagup | \diagdown F \\ F \end{array}$$

C
$$\begin{array}{c} F \\ F \diagdown | \diagup F \\ \quad Cr \\ H_2O \diagup | \diagdown F \\ F \end{array}$$

D
$$\begin{array}{c} F \\ H_2O \diagdown | \diagup F \\ \quad Cr \\ F \diagup | \diagdown F \\ H_2O \end{array}$$

E
$$\begin{array}{c} H_2O \\ H_2O \diagdown | \diagup F \\ \quad Cr \\ F \diagup | \diagdown F \\ F \end{array}$$

10 How many different isomeric forms are there of the complex ion

$CrF_5(H_2O)^{2-}$

assuming it has a similar shape to the complex shown in item 9?

A 1   B 2   C 3   D 4   E 6

11-13 These items concern the following stability constants for the formation of iron(III) complexes from hydrated iron(III) ions:

Ligand	lg K_1	lg K_2	lg K_3	lg K_4	lg K_5	lg K
CNS⁻ (thiocyanate)	2.0	1.4	0.8	0.02		4.2
F⁻ (fluoride)	5.3	4.5	3.2	2.0	0.36	15.4
edta	25.1					25.1

11 Why is only one figure given for the edta complex?

A Because the hydrated iron(III) ion has the formula $Fe(H_2O)_4^{3+}$ and therefore one edta molecule has to be shared between two iron(III) ions
B Because the ion Fe^{3+}(aq) has the formula $Fe(H_2O)_4^+$ and edta is quadridentate
C Because only one water molecule in the hydrated iron(III) ion is replaced by an edta molecule
D Because edta occupies all four coordination positions in a tetrahedral complex so only one molecule is necessary
E Because one edta molecule has the ability to take the place of six water molecules in a complex ion

12 Which of the following best describes the situation when thiocyanate ions are added to a solution of iron(III) ions in aqueous sodium fluoride?

- **A** One fluoride ion is replaced by a thiocyanate ion but the reaction does not proceed further
- **B** Five thiocyanate ions replace the appropriate number of fluoride ions
- **C** Unless the concentration of thiocyanate ions is very high, there is no appreciable reaction
- **D** Because of the low stability constants of both the thiocyanate and the fluoride complexes, neither ion complexes iron(III) ions
- **E** The thiocyanate ions replace the fluoride ions completely giving an extremely stable complex

13 The edta complex with copper(II) ions has lg K = 18.8. Which of the following consequences would be expected?

- **A** If aqueous iron(III) ions are added to a solution of copper(II) ions in aqueous edta, the edta would transfer to the iron ions
- **B** If aqueous copper(II) ions are added to a solution of iron(III) ions in edta, the edta would transfer to the copper ions
- **C** If aqueous edta were added to an equimolar mixture of copper(II) and iron(III) ions, the edta would divide itself between the two in the ratio 18.8 to 25.1
- **D** If aqueous edta were added to a solution of copper(II) ions the edta would form complexes with $\frac{18.8}{19.8} \times 100\%$ of the copper ions
- **E** If aqueous edta were added to an equimolar solution of copper(II) and iron(III) ions, it is impossible to predict what would happen to the edta because the iron(III) ions would oxidize the copper(II) ions

14 All of the following statements about zinc are true; in which way is zinc behaving untypically for a transition element?

1. The ion of zinc has a full d shell
2. The compounds of zinc are colourless
3. In its compounds zinc has only one oxidation state

15-20 These items concern the information contained in the diagram on page 75.

15 In the equations accompanying each $E^{\ominus}$ value and oxidation number change the reducing agents are to be found:

- **A** on the right of the equations
- **B** on the left of the equations
- **C** on the right for $E^{\ominus}$ values which are positive
- **D** on the left for $E^{\ominus}$ values which are positive
- **E** on the left if acid solutions are being used, on the right otherwise

16 According to the information provided on the diagram, which of the following would you expect to liberate hydrogen from dilute hydrochloric acid?

A $Cr^{3+}(aq)$ **B** $I^-(aq)$ **C** $Cu(s)$ **D** $Cr(OH)_3(s)$ **E** $Cr(s)$

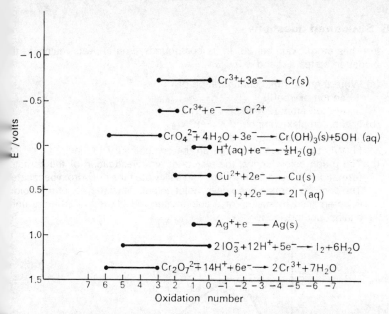

7 Which of the following would you expect to be able to oxidize iodine to a higher oxidation state?

- **A** Cr^{3+}(aq) in acid solution
- **B** CrO_4^{2-}(aq) in alkaline solution
- **C** $Cr_2O_7^{2-}$(aq) in acid solution
- **D** Ag^+(aq)
- **E** It is not possible to say from the information supplied

8 Which of the following would you expect to be able to oxidize iodide ions most rapidly to iodine?

- **A** Cu(s)
- **B** $Cr_2O_7^{2-}$(aq)
- **C** IO_3^-(aq)
- **D** CrO_4^{2-}(aq)
- **E** It is not possible to say from the information supplied

19 Which of the following would be the most likely observation if a solution of silver ions were added to a solution containing iodide ions?

- **A** There would be a deposit of silver on the inside of the test tube
- **B** The solution would turn a very dark brown
- **C** The solution would turn a very dark brown, then slowly colourless
- **D** A pale yellow precipitate would be formed
- **E** The solution would turn purple

20 Suppose that some of a solution containing Cr^{3+} ions were added to some of a solution containing iodide ions. Which of the following would be expected to occur?

- **A** Iodine would be produced immediately
- **B** Iodine would be produced very slowly
- **C** Cr^{3+} would be reduced to Cr
- **D** Cr^{3+} would be reduced to Cr^{2+}
- **E** Nothing would happen

16B Structured questions

1. Iron has atomic number 26. In its compounds, iron appears mainly in the oxidation states +2 and +3.

   (a) Write the electronic configuration of
      (i) an ion of iron(II)
      (ii) an iron atom (2 marks)

   (b) In the complex compound $K_3Fe(CN)_6$
      (i) What is the shape of the complex ion? Sketch it.
      (ii) What is the overall charge on the complex ion? (3 marks)

   (c) The graph below shows the results of an investigation of the complex formation between iron(III) ions and salicylate ions (2-hydroxybenzoate). The complex formed has a deep violet colour. The intensity of this colour is determined by means of a colorimeter and plotted in arbitrary units against composition.

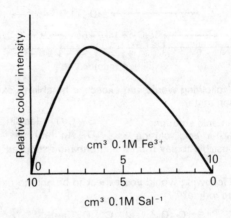

      (i) The composition of the complex may be represented by $Fe_x(Sal)_y$. From the graph, what are the most likely values for x and y?
      (ii) The charge on the salicylate ion is -1. What is the overall charge on the complex?
      (iii) Write the expression for the overall stability constant of the complex. (5 marks)

2. The chart on page 77 shows some redox potentials of rhenium (Re) and its compounds.

   (a) What compound or ion not containing rhenium could change ReO_4^- to ReO_2? (1 mark)

   (b) What compound or ion not containing rhenium could change ReO_4^- to Re? (1 mark)

   (c) (i) What reaction is possible between MnO_2 and ReO_2?
      (ii) Why might this reaction not take place in practice? (3 marks)

   (d) (i) What reaction is possible between ReO_4^- and Cu?
      (ii) What would be the effect of increasing the $[H^+]$ on the outcome of this reaction? (3 marks)

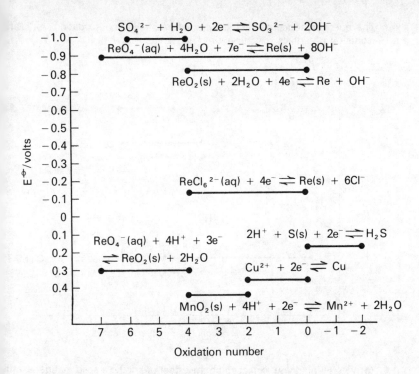

(e) Sketch a practical arrangement which could be used for measuring $E^{\ominus}$ for the ReO_4^-/Re electrode. Label the diagram fully but do not include the hydrogen electrode. (2 marks)

3. Platinum ions Pt^{2+} are capable of forming complexes with ammonia NH_3, 1,2-diaminoethane $NH_2CH_2CH_2NH_2$, Cl^- and OH^-. Indeed, two types of ligand may complex with the same platinum ion at the same time. All the complexes are square planar in shape.

(a) What is meant by the expression 'bidentate ligand'? Give one example of such a ligand from the list above. (2 marks)

(b) It is possible to prepare *two* complexes of formula $Pt(NH_3)_2Cl_2$. Sketch the arrangement of the ligands in both these complexes. (2 marks)

(c) One of the two complexes mentioned in (ii) will react with 1,2-diaminoethane (sometimes abbreviated to 'en') to give a complex of formula

   $[Pt(NH_3)_2en]^{n+}$

   What is the value of n? (2 marks)

(d) Which of the two complexes mentioned in (b) is the more likely to react with diaminoethane? Justify your answer. (2 marks)

(e) When either of the complexes mentioned in (b) is reacted with a mixture of silver oxide and water (which behaves as 'AgOH'), the by-product is two moles of silver chloride AgCl per mole of complex. What is the likely formula of the platinum complex that results? (2 marks)

4 This question concerns the following information about oxidation states and oxidation potentials for chromium and manganese

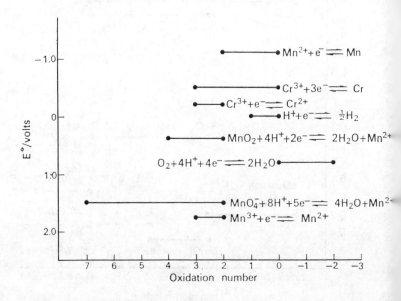

(a) What would you expect to be the effect of adding a solid manganese(III) compound to water? (2 marks)
(b) Suggest a means whereby the +3 oxidation state of manganese could be stabilized in solution. (1 mark)
(c) Which is the more powerful reducing agent, Mn^{2+}(aq) or Cr^{2+}(aq)? Justify your answer. (2 marks)
(d) What would you expect to be the effect of adding an acidified solution of potassium manganate(VII) $KMnO_4$ to
  (i) a solution of Mn^{2+}(aq)?
  (ii) a solution of Cr^{2+}(aq)? (3 marks)
(e) What would you expect to be the effect of attempting to react
  (i) manganese metal
  (ii) chromium metal
  with dilute hydrochloric acid? (2 marks)

5 The free-energy changes associated with the formation of various oxides are represented on the diagram on page 79. Use this diagram to answer the questions.

(a) The standard free-energy change $\Delta G^{\ominus}$ is calculated from the relationship $\Delta G^{\ominus} = RT \ln p_{O_2}$. A more usual relationship would be $\Delta G^{\ominus} = -RT \ln k_p$. Why is it legitimate to use the first of these expressions for most of the $\Delta G^{\ominus}$ values? (2 marks)
(b) In which cases is it incorrect to use this expression? (2 marks)
(c) What substances would be present in the resulting mixture if carbon were allowed to reach equilibrium with oxygen at 1100 K? (2 marks)

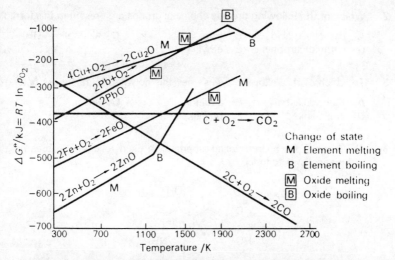

(d) At what temperature is the reaction between carbon and zinc oxide such that the equilibrium constant for the reaction is 1? (1 mark)
(e) What happens to the composition of the reaction mixture in the reaction between zinc oxide and carbon if the temperature is raised? (1 mark)
(f) Write the chemical equation for this same reaction, incorporating the correct state symbols for reactants and products. (2 marks)

SECTION 17

Organic chemistry 4

17A Fixed response items

1-3 These items concern a polymer with the structure

$$\begin{array}{c} CH_3 \ \ \ H \ \ \ \ \ \ \ \ CH_3 \ \ \ H \ \ \ \ \ \ \ \ CH_3 \ \ \ H \\ | \ \ \ \ | \ \ \ \ \ \ \ \ \ \ \ | \ \ \ \ | \ \ \ \ \ \ \ \ \ \ \ | \ \ \ \ | \\ -C-C-\ \ \ \ \ \ \ -C-C-\ \ \ \ \ \ \ -C-C- \\ | \ \ \ \ | \ \ \ \ \ \ \ \ \ \ \ | \ \ \ \ | \ \ \ \ \ \ \ \ \ \ \ | \ \ \ \ | \\ H \ \ \ CO_2CH_3 \ \ \ H \ \ \ CO_2CH_3 \ \ \ H \ \ \ CO_2CH_3 \end{array}$$

1 Which of the following might have been the monomer from which this polymer was made?

A
$$\begin{array}{c} CH_3 \ \ \ H \\ | \ \ \ \ | \\ C=C \\ | \ \ \ \ | \\ H \ \ \ CO_2CH_3 \end{array}$$

B
$$\begin{array}{c} CH_3 \ \ \ H \\ | \ \ \ \ | \\ H-C-C-H \\ | \ \ \ \ | \\ H \ \ \ CO_2CH_3 \end{array}$$

C
$$\begin{array}{c} CH_3 \ \ \ \ \ \ \ \ \ \ \ \ H \\ | \ \ \ \ \ \ \ \ \ \ \ \ \ \ / \\ C=C \\ | \ \ \ \ \ \ \ \ \ \ \ \ \ \ \backslash \\ CO_2CH_3 \ \ \ \ H \end{array}$$

D
$$\begin{array}{c} CH_3 \ \ \ H \\ | \ \ \ \ | \\ H-C-C-CH_3 \\ | \ \ \ \ | \\ H \ \ \ CO_2H \end{array}$$

E
$$\begin{array}{c} CH_3 \ \ \ H \\ | \ \ \ \ | \\ CH_3-C-C-CH_3 \\ | \ \ \ \ | \\ H \ \ \ CO_2CH_3 \end{array}$$

2   Which of the following organic chemical groups are present in this formula
    A carboxylic acid   B ester   C alcoholic $-OH$
    D ketonic carbonyl   E ethyl

3   To which of the following types does this polymer definitely *not* belong?
    A thermoplastic   B addition   C linear
    D cross-linked   E polyalkene

4   The formula for a drug called isoprenaline used to ease the breathing of asthmatics is as follows:

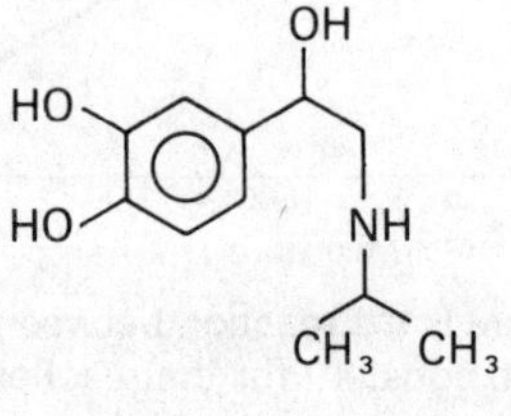

    [structure: benzene ring with two HO− groups, and a −CH(OH)−CH$_2$−NH−CH(CH$_3$)$_2$ side chain]

    In which of the following ways is isoprenaline structurally similar to aspirin?
    1 Both contain alcoholic $-OH$
    2 Both are secondary amines
    3 Both contain a benzene ring

5   A substance called 'oil of wintergreen' is used in various kinds of liniment. It has the formula

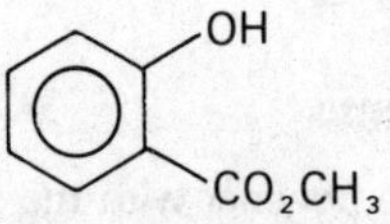

    [structure: benzene ring with −OH and −CO$_2$CH$_3$ groups in ortho positions]

    Starting from 2-hydroxybenzoic acid, reaction with which of the following reactants would give oil of wintergreen?

    A sodium hydroxide and then methanol
    B methane at 150°C
    C chloromethane
    D methylamine
    E methanol and conc. sulphuric acid

6   By which of the following reactions could 2-hydroxychlorobenzene be made from 2-aminochlorobenzene?

    A reflux with water
    B reflux with dilute hydrochloric acid
    C reflux with sodium hydroxide solution
    D diazotize and then warm the mixture
    E diazotize and then add ethanol

7-10 These items concern the biosynthesis of adrenaline. This takes the following route, each stage of which is catalysed by an enzyme:

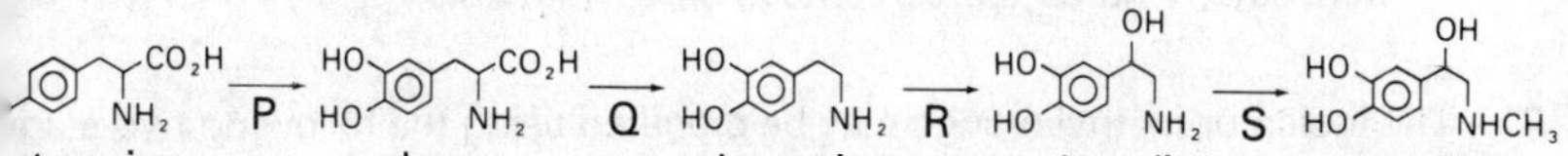

tyrosine →P→ dopa →Q→ dopamine →R→ noradrenaline →S→ adrenaline

7 Which of the five substances is *not* chiral?

A tyrosine
B dopa
C dopamine
D noradrenaline
E adrenaline

8 In which stage is an alcoholic −OH group introduced?

A P   B Q   C R   D S
E None of these stages

9 Which of the following best describes the type of reaction which is most likely to be happening in stage R?

A oxidation   B reduction   C hydrolysis
D nucleophilic attack   E electrophilic attack

10 If stage S were to be attempted in the laboratory which of the following reagents would be most likely to succeed?

A methanol   B methane   C methylamine
D methanal   E bromomethane

17B Structured questions

1 The questions which follow concern this synthetic step:

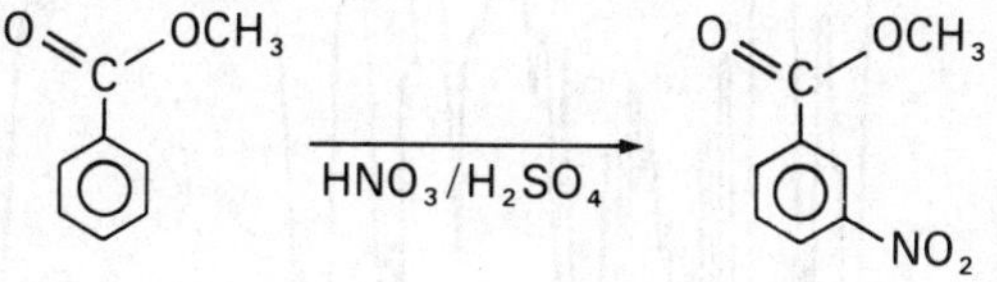

methylbenzoate (RMM = 120) →[HNO₃/H₂SO₄]→ methyl-3-nitrobenzoate (RMM = 165)

(a) What is the effective nitrating agent in this nitration? (2 marks)
(b) Is the reactant mentioned in (a) an electrophile or a nucleophile? (1 mark)
(c) Explain how to carry out the recrystallization of the crude product from ethanol. (3 marks)

(d) After recrystallization how, in principle, could you confirm the purity of your product? (Practical detail is *not* required.) (1 marks)

(e) If 1.85 g of methyl-3-nitrobenzoate is produced from 1.5 g of methyl benzoate, what is the percentage yield? (3 marks)

2 The indicator methyl orange may be prepared using the following two-stage synthesis:

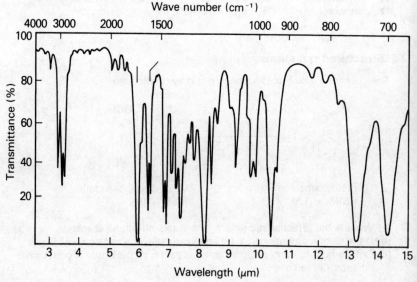

4-aminobenzenesulphonic acid   methyl orange

(a) How would nitrous acid be obtained for the first step? (2 marks)
(b) What temperature conditions would be appropriate for the first step? (1 mark)
(c) What colour would you expect the mixture to have after the second step? Justify your answer. (2 marks)
(d) In the acid conditions used in the first step, the 4-aminobenzenesulphonic acid would have accepted a proton. Write the formula for the ion resulting from this. (2 marks)
(e) If the nitrous acid is added too quickly, a colourless gas is evolved. What is this gas? (1 mark)
(f) Methyl orange, like many indicators, is a weak acid. Write the formula for the ion which results when a molecule of methyl orange loses a proton. (2 marks)

3 A compound of molecular formula $C_9H_{10}O$ has an infra-red spectrum as follows:

Characteristic infra-red absorptions for a variety of organic molecules[12]

Molecule or group	Vibration type	Wave number/cm^{-1}
Alkyl group (CH$_3$, CH$_2$, CH)	C–H stretch	2960–2850
	C–H bend	1460–1370
Alkanal (CHO)	C–H stretch	2900–2700
Alkyne (C≡CH)	C–H stretch	3300–3270
Alkene (C=CH$_2$)	C–H stretch	3905–3075
	C–H bend	990–890
Arene	C–H stretch	3040–3010
	C–H bend: in-plane	1300–1000
	out-of-plane	900–650
Alkanol (OH)	O–H stretch	3650–3590
	C–O stretch	1200–1050
Amine, amide (NH$_2$)	N–H stretch	3500–3300
Aliphatic ketone (R$_2$CO)	C=O stretch	1740–1700
Aliphatic alkanal (RCHO)	C=O stretch	1740–1720
Aromatic ketone (Ar$_2$CO)	C=O stretch	1700–1680
Alkanoic acid (RCO$_2$H)	C=O stretch	1725–1700
Alkanoyl chloride (RCOCl)	C=O stretch	1815–1790
Alkanoate ester (RCO$_2$R')	C=O stretch	1750–1730
	C–O stretch	1300–1050
Alkoxy (ether) R$_2$O	C–O stretch	1150–1070

(a) What is 'wave number' a measure of? (1 mark)
(b) What is the significance of the two prominent peaks at wave numbers less than 800 cm^{-1}? (1 mark)
(c) What causes the prominent peak at just over 1700 cm^{-1}? (1 mark)
(d) Explain why the spectrum shows that the compound is not an alcohol. (1 mark)
(e) The peaks in the region of 3000 cm^{-1} are caused by vibrations involving C–H bonds. Suggest a reason why there are several peaks close together. (2 marks)
(f) Suggest a structural formula for a compound which fits the evidence given by the spectrum. (2 marks)
(g) Write the formula for *one* other isomer with the molecular formula C$_9$H$_{10}$O and mention one way in which the spectrum of your isomer would be different from the spectrum illustrated. (2 marks)

4 A compound whose molecule contains only carbon, hydrogen and bromine atoms has the mass spectrum shown.

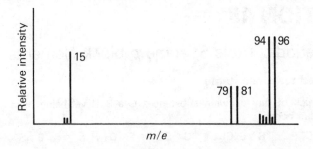

(a) What does the ratio m/e mean? (2 marks)
(b) What causes the peak at $m/e = 15$? (1 mark)
(c) What is the significance of the two pairs of peaks at m/e values of 79/8 and 94/96? (2 marks)
(d) What is the formula for the compound? (1 mark)
(e) Why are there small peaks at m/e values of 91, 92, 93 and 95? (2 mark
(f) If the charges on the ions which pass down the mass spectrometer wer doubled, write the m/e value for any *one* new peak which would t present on the mass spectrum and give the formula for the ic concerned. (2 marks)

5   A hydrocarbon X contains a benzene ring and gives a mass spectrum a shown below. Vigorous oxidation of X produces benzene-1,4-dicarboxyli acid.

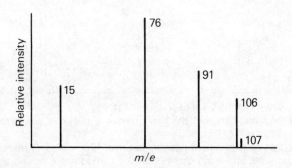

(a) What is the relative molecular mass of X? (1 mark)
(b) What are the species most probably responsible for the fragmentatior peaks at m/e values of 76, 91 and 106? (3 marks)
(c) Write down the formulae of 3 isomers of X, indicating which of them is X itself. (3 marks)
(d) In a reaction to produce X the theoretical yield was 0.1 mole but the actual yield was 7.7 g. What was the percentage yield of X? (2 marks)
(e) What is responsible for the small peak at $m/e = 107$? (1 mark)

SECTION 18

The Periodic Table 5: some *p*-block elements

18A Fixed response items

1   In which of the following substances is a 3d level being used to form covalent bonds?

   A SCl_2 B S_2Cl_2 C SO_2 D H_2S E S_8

In which of the following is sulphur forming four covalent bonds?

A SO_3^{2-} B SO_3 C H_2SO_4 D H_2S E $S_2O_3^{2-}$

In which of the following is delocalization of electrons most likely to take place?

A $O=S=O$

B $\begin{array}{c} H-O \\ H-O \end{array} S \begin{array}{c} O \\ O \end{array}$

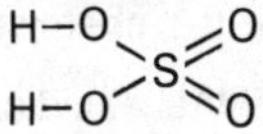

C $H-O-N=O$

D $^-O-S \begin{array}{c} O^- \\ O \end{array}$

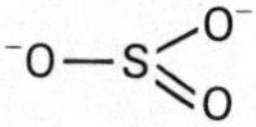

E S^{2-}

In which of the following compounds of nitrogen does the nitrogen atom not have the electronic configuration of a noble gas?

1 NO_2 i.e. $O=N \rightarrow O$
2 NO i.e. $N=O$
3 N_2O_4 i.e. $\begin{array}{c} O \\ O \end{array} N-N \begin{array}{c} O \\ O \end{array}$

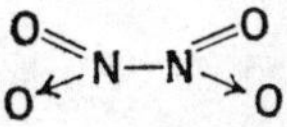

5-9 Oxides may be classified under the following headings

A acidic — react with alkalis
B basic — react with acids
C amphoteric — react with both acids and alkalis
D neutral — react with neither acids nor alkalis
E peroxides — contain the ion O_2^{2-}

Choose the heading from the above list which is most appropriate for each of the oxides mentioned in items 5 to 8.

5 dinitrogen oxide, N_2O

6 aluminium oxide, Al_2O_3

7 tin(IV) oxide, SnO_2

8 nitrogen monoxide, NO

9 the oxide of barium, BaO_2

10 Which of the following statements about boron compounds is/are true?
1 boron has an oxidation number of $+3$ in its oxides and halides
2 the compound of formula $B(OH)_3$ is acidic in character
3 the chloride of boron is readily hydrolysed by water

11 When lead(IV) oxide reacts with hot dilute hydrochloric acid

1. chlorine is formed
2. lead(II) chloride is formed
3. oxygen is formed

12 Gallium is the element below aluminium in Group III. Which of the following expectations is most likely for the chemistry of gallium compounds?

A gallium oxide would be entirely acidic
B gallium chloride would hydrolyse in water
C gallium chloride would be an oxidising agent
D gallium hydroxide would have the formula $Ga(OH)_2$
E gallium chloride would be a reducing agent

13-14 These items refer to the following information:

For the half equation

$$SO_2(g) + 5H_2O(l) \rightarrow HSO_4^-(aq) + 3H_3O^+(aq) + 2e^-$$

the redox potential is -0.11 volts.

13 Written conventionally, the electrode diagram for this process would be:

A $[SO_2(g) + 5H_2O(l)], [HSO_4^-(aq) + 3H_3O^+(aq)] \mid Pt \quad E^\ominus = -0.11$ V
B $SO_2(g), HSO_4^-(aq) \mid Pt \quad E^\ominus = -0.11$ V
C $[HSO_4^-(aq) + 3H_3O^+(aq)], [SO_2(g) + 5H_2O(l)] \mid Pt \quad E^\ominus = +0.11$ V
D $HSO_4^-(aq) + H_3O^+(aq) \mid SO_2(g) + 5H_2O(l) \quad E^\ominus = +0.11$ V
E $SO_2(g) + 5H_2O(l) \mid HSO_4^-(aq) + 3H_3O^+(aq) \quad E^\ominus = -0.11$ V

14 Which of the following diagrams would be most appropriate for this reaction? ($E^\ominus$ is on the vertical axis, oxidation number of sulphur on the horizontal axis)

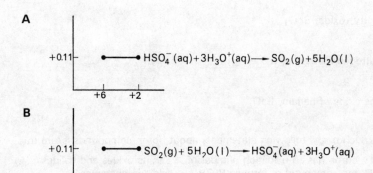

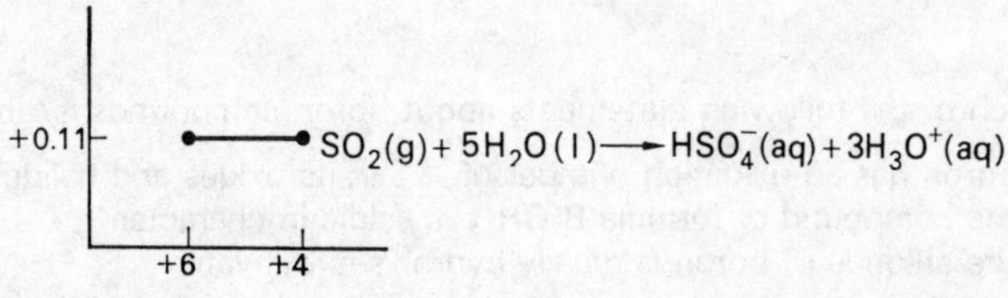

C

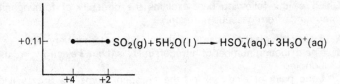

D

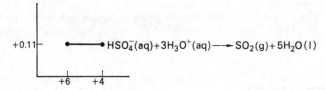

E

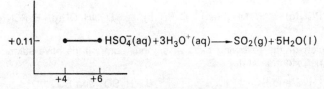

5 When persulphate ions ($S_2O_8^{2-}$) are reduced to sulphate ions (SO_4^{2-}), which of the following best represents the electronic changes which take place?

A $S_2O_8^{2-}(aq) + 2e^- \rightleftharpoons 2SO_4^{2-}(aq)$
B $S_2O_8^{2-}(aq) \rightarrow 2SO_4^{2-}(aq) + 2e^-$
C $S_2O_8^{2-}(aq) + e^- \rightarrow 2SO_4^{2-}(aq)$
D $S_2O_8^{2-}(aq) \rightarrow 2SO_4^{2-}(aq) + e^-$
E $S_2O_8^{2-}(aq) + 4e^- \rightarrow 2SO_4^{2-}(aq)$

6 Which of the following reactions could be described as a 'disproportionation'?

A $SO_2(g) + H_2O(l) \rightleftharpoons HSO_3^-(aq) + H^+(aq)$
B $2SO_2(aq) + 2H^+(aq) + 4e^- \rightarrow S_2O_3^{2-}(aq) + H_2O(l)$
C $3Cu(s) + 12H^+(aq) + 6NO_3^-(aq) \rightarrow 2N_2O_4(g) + 6H_2O(l) + 3Cu^{2+}(aq)$
D $3HSO_3^-(aq) + OH^-(aq) \rightarrow S(s) + 2SO_4^{2-}(aq) + 2H_2O(l)$
E $NH_4NO_3(s) \rightarrow N_2O(g) + 2H_2O(l)$

7 The standard redox potential for the electrode

[$2NO_3^-(aq) + 10H^+(aq)$], [$N_2O(g) + 5H_2O(l)$] | Pt

is +1.1 V at pH 0. Which of the following statements is/are correct?

1 This redox potential is measured against the standard hydrogen electrode
2 The e.m.f. of this electrode would vary with change in pH
3 The oxidation state of nitrogen in one of these nitrogen compounds is −1

18   Which of the following best explains the difficulty of forming nitrogen compounds from gaseous nitrogen?

   A   The first ionization energy of the nitrogen atom is very high
   B   The oxidation number of nitrogen in N_2 is 0 but 3 electrons are used for bonding
   C   Lone pairs of electrons on the nitrogen atoms are not available for bonding
   D   The double bond in nitrogen is twice as strong as a single N—N bond
   E   The bond dissociation energy of N_2 is very high, giving a very stable molecule

19   In which of the following compounds is a 3d electron level being used for covalent bonding?

   A $PCl_5(g)$ B $NH_4^+(aq)$ C $PH_3(g)$ D $NH_2OH(g)$ E $PCl_3(g)$

20   Which of the following pairs of sulphur compounds have sulphur in the same oxidation state?

   A H_2S and SCl_2
   B H_2SO_4 and SO_3
   C $Na_2S_2O_8$ and $Na_2S_4O_6$
   D Na_2SO_3 and $Na_2S_2O_6$
   E $Na_2S_2O_4$ and $Na_2S_4O_6$

18B Structured questions

1   Tin(IV) iodide may be prepared by refluxing 2 g of granulated tin with a solution of 6 g of iodine in 25 cm³ of a suitable non-aqueous solvent. The reaction may need a little heat to start it but is exothermic so further heating is unnecessary.

   (a) Name a suitable non-aqueous solvent to use in this preparation. (1 mark)
   (b) What colour would you expect the mixture to be during the reaction? (1 mark)
   (c) Show by calculation and the writing of an equation whether the tin or the iodine is in excess. (3 marks)
   (d) How would you know when the reaction is complete or has reached equilibrium? (2 marks)
   (e) If some iodine remained unreacted, what might you add to the mixture to discharge the colour of the iodine? (1 mark)
   On cooling, tin(IV) iodide crystallizes as orange crystals, melting at 144°C
   (f) Describe how the melting point of the crystals could be measured (2 marks)

2   The questions which follow concern some compounds of nitrogen, notably hydrazine N_2H_4. Hydrazine can be made by a controlled reaction between ammonia and chlorine

   $Cl_2(g) + 4NH_3(aq) \rightarrow N_2H_4(aq) + 2NH_4^+(aq) + 2Cl^-(aq)$

   When pure, hydrazine is a colourless liquid at room temperature. It is unstable with respect to decomposition to nitrogen, ammonia and hydrogen. It is a base.

(a) What is the oxidation number of nitrogen in hydrazine? (2 marks)
(b) What is the name given to the type of redox reaction by which hydrazine decomposes? (1 mark)
(c) Write a balanced equation for this reaction. (2 marks)
(d) Hydrazine is a liquid which is miscible with water but explosive when heated in air. Suggest a method of obtaining a sample of hydrazine from the reaction mixture resulting from its preparation. (2 marks)
(e) Hydrazine is said to be a base. Write an ionic equation showing a reaction of hydrazine with H_3O^+ ions. How many such equations can be written? (3 marks)

Redox potentials at pH = 0 are shown on the following chart:

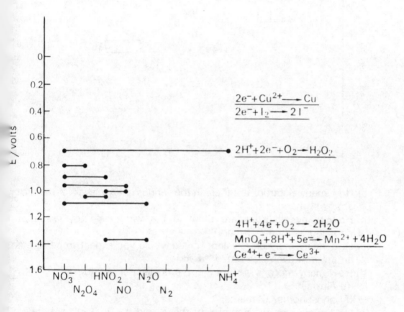

(a) Redraw part of the above diagram so as to show *one* example of an opportunity for disproportionation to occur. (2 marks)
(b) As may be seen from the diagram, elemental nitrogen does not participate in reactions with aqueous solutions. What is the principal reason for this lack of reactivity? (2 marks)
(c) Give an example of a reducing agent that might be able to reduce nitrous acid HNO_2 to dinitrogen oxide N_2O. (1 mark)
(d) What are the possible nitrogen-containing products of the action of cerium(IV) ion Ce^{4+} on nitrous acid HNO_2? (2 marks)
(e) Given a standard solution of cerium(IV) ions and a standard solution of acidified sodium nitrite (nitrous acid), explain briefly how you would attempt to find out which of the possible nitrogen-containing products is in fact formed. (2 marks)
(f) Why is it that in the diagram nitrous acid is represented by HNO_2 whereas nitric acid is shown as its ion NO_3^-? (1 mark)

4 The electronic energy levels of the nitrogen and phosphorus atoms shown on the following diagram:

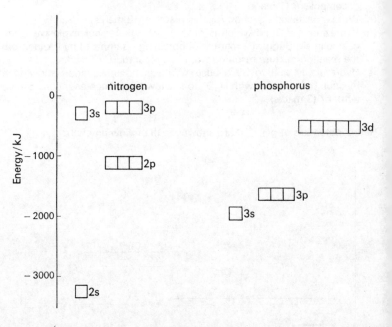

(a) How many electrons are there in the various energy levels of the nitrogen atom? (1 mark)
(b) How many electrons are there in the various energy levels of the phosphorus atom? (1 mark)
(c) How many 'normal' covalent bonds would you expect to be formed an atom of each element? (1 mark)
(d) How many protons are there in the nucleus of an atom of
  (i) nitrogen
  (ii) phosphorus? (2 marks)
(e) On the diagram the energies of the 3s and the 3p levels are shown both nitrogen and phosphorus. Why are the energies of these leve greater for nitrogen than for phosphorus? (2 marks)
(f) Remembering that bond energies for covalent bonds tend to be of t order of 400–800 kJ mol^{-1}, briefly explain why PCl$_3$ and PCl$_5$ molecu both exist whereas NCl$_3$ exists but not NCl$_5$. (3 marks)